CW01670023

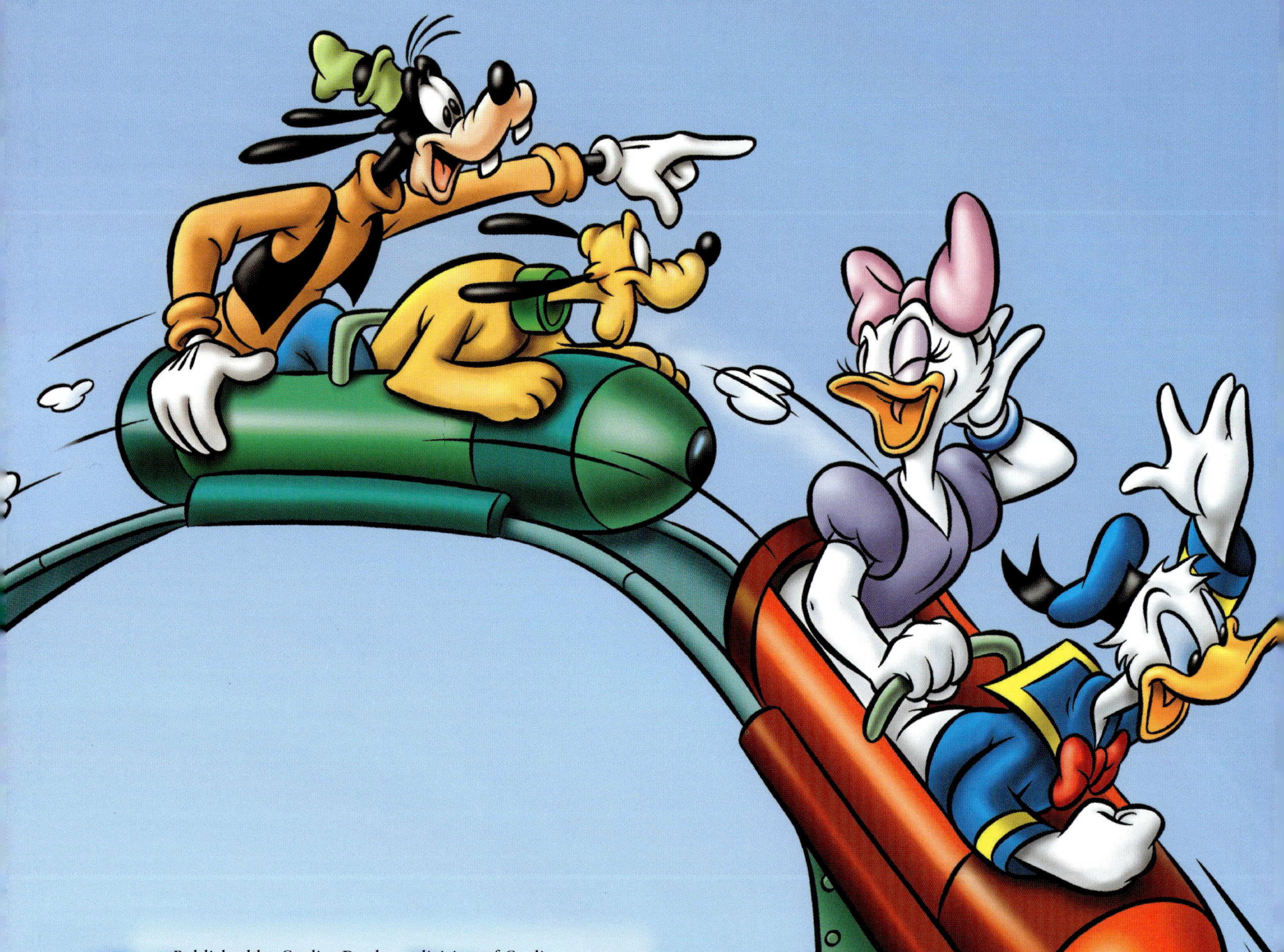

Published by Grolier Books, a division of Grolier Enterprises Inc.

Disney Presents The Wonderful World of Knowledge
ISBN 0-7172-8929-X
Planet Earth ISBN 0-7172-8945-1

© 1999 Disney

First published in 1999

All rights reserved. No part of this book may be reproduced or transmitted in any form by any means, electronic, mechanical, or otherwise, whether now or hereafter devised, including photocopying, recording, or by any information storage and retrieval system without express written prior permission from the publisher.

Printed and bound in China by
Toppan Printing Company

Originated in Italy by Articolor

Designed and compiled by
Marshall Editions Developments Limited

GROLIER BOOKS

Disney PRESENTS

The Wonderful World of Knowledge

PLANET EARTH

Using The Wonderful World of Knowledge

Mickey, Minnie, Donald, Daisy, Goofy, and Pluto are ready to take you on an adventure ride through the world of learning. Discover the secrets of science, nature, our world, the past, and much more. Climb aboard and enjoy the ride.

Look here for a general summary of the theme

Labels tell *you what's happening in the pictures*

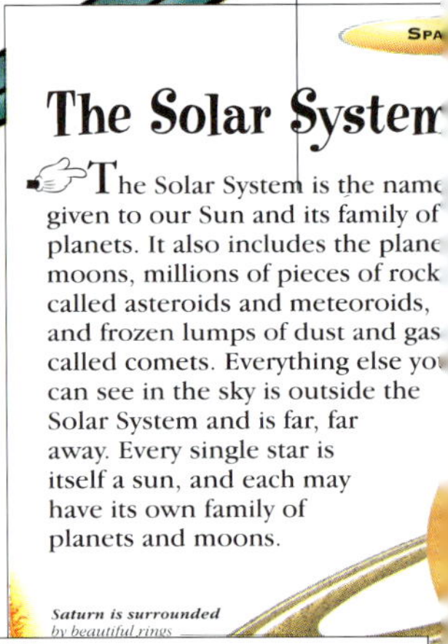

Mickey's ears *lead you to one of the main topics*

The pictures by *themselves can tell you a lot, even before you read a word*

Watch out for special pages where Mickey takes a close look at some key ideas

Mickey's page *numbers help you look things up. Don't forget there's a glossary and index at the back of each book*

Goofy and his *friends know how to give you a chuckle on every topic*

Mickey points you to more information in other books in your *The Wonderful World of Knowledge*

FIND OUT MORE
MAMMALS: Camouflage
PLANET EARTH: Forests

Amazing Facts

★ The Sun is enormous compared to the planets. It is nearly 1,000 times more massive than the giant planet Jupiter.

Your favorite characters present some facts to astound you and your friends

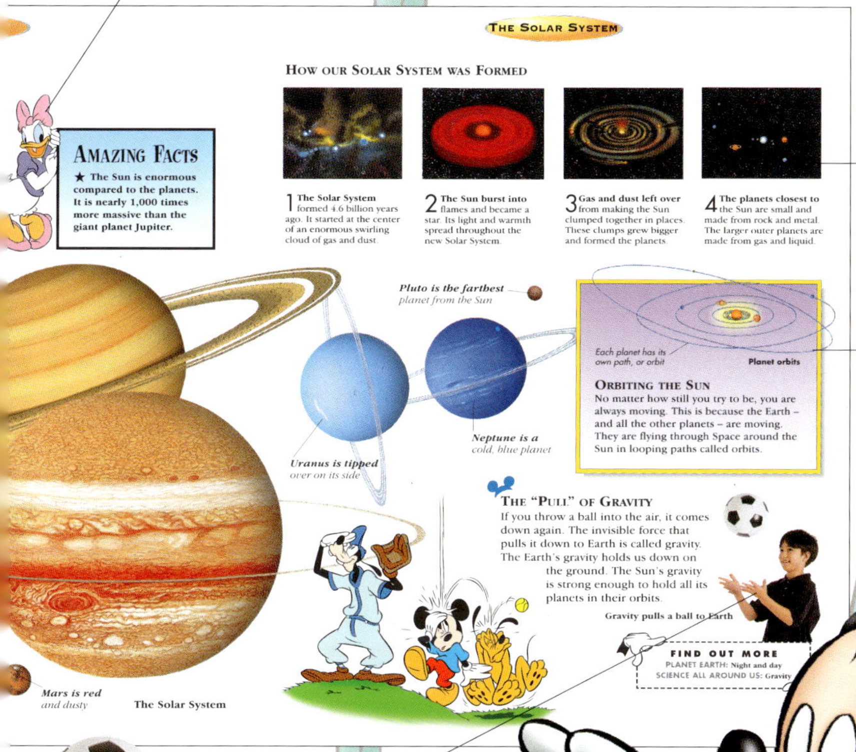

The Solar System

How our Solar System was Formed

1 The Solar System formed 4.6 billion years ago. It started at the center of an enormous swirling cloud of gas and dust.

2 The Sun burst into flames and became a star. Its light and warmth spread throughout the new Solar System.

3 Gas and dust left over from making the Sun clumped together in places. These clumps grew bigger and formed the planets.

4 The planets closest to the Sun are small and made from rock and metal. The larger outer planets are made from gas and liquid.

Amazing Facts

★ The Sun is enormous compared to the planets. It is nearly 1,000 times more massive than the giant planet Jupiter.

The Solar System

Orbiting the Sun

No matter how still you try to be, you are always moving. This is because the Earth – and all the other planets – are moving. They are flying through Space around the Sun in looping paths called orbits.

The "Pull" of Gravity

If you throw a ball into the air, it comes down again. The invisible force that pulls it down to Earth is called gravity. The Earth's gravity holds us down on the ground. The Sun's gravity is strong enough to hold all its planets in their orbits.

Gravity pulls a ball to Earth

FIND OUT MORE
PLANET EARTH: Night and day
SCIENCE ALL AROUND US: Gravity

Numbers lead *you step-by-step through how things happen*

Colorful boxes *zoom in on information*

Mickey's helpers test some ideas themselves

Contents

INTRODUCING

Planet Earth

Millions of years ago, a huge ball of very hot rock spun around the Sun. Gradually, it cooled down. When a crust formed, hot gases burst through it, turned into air, and clouds and water formed. Finally, living things appeared. Our Earth had begun its life.

Planet Earth has settled into the landscapes we know, such as forests, deserts, and grasslands. Volcanoes and earthquakes, hurricanes and tornadoes show the power that went into its creation. And even today our planet is being changed by the forces of wind, ice, and water.

The Story of the Earth

Planet Earth is a huge ball of rock that spins around a star called the Sun. The Earth was formed about 4.6 billion years ago from hot gas and dust, which then became a ball of hot, liquid rock. Over millions of years, the temperature at the center of the ball cooled down to about 5,500°C (9,932°F). Gases escaped from deep inside the Earth and built up around it, forming a bubble of gas called the atmosphere.

AN ASTRONAUT'S VIEW

Seen from the Moon, the Earth looks round, but really it is slightly squashed at the top and bottom, and bulges out around the middle. A lot of the Earth is covered in water, making the planet look blue.

Earth seen from the Moon

HOW THE EARTH FORMED

1 **In the early years of the Earth's story,** the planet was a ball of hot, liquid rock. As its surface formed, gases burst out through volcanoes and formed an atmosphere around the planet.

2 **The Earth gradually cooled down, and the** liquid rock on the surface set solid. Clouds formed in the new atmosphere and it began to rain. The oceans and seas started to fill with water.

THE SOLAR SYSTEM

The Earth spins around the Sun with eight other planets in a part of Space called the Solar System. It is the only planet in the Solar System where life is known to exist. Our Solar System is part of a group of stars and planets called the Milky Way.

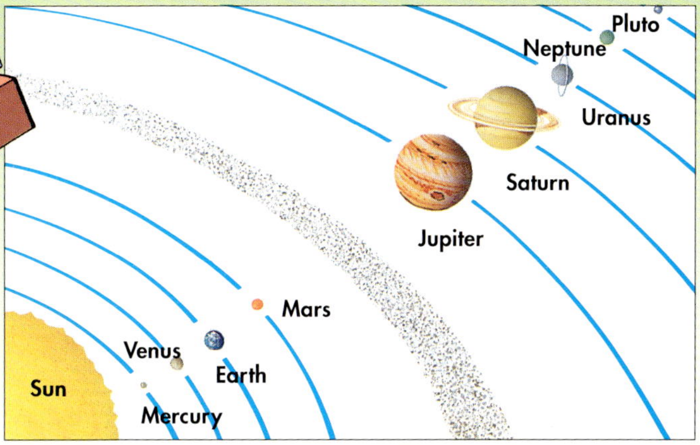

3 About 3 billion years ago, the first life appeared in the oceans. At first there were only simple life forms. Then plants grew, animals developed, and life moved from the water to the land.

4 Today there are many different plants and animals on the Earth. The planet is still cooling down and is very hot in the center. There are still volcanoes and storms, but not as many as there were when the Earth was being formed.

AMAZING FACTS

★ **The Sun is so large that over 1 million Earths would fit inside it.**

FIND OUT MORE
DINOSAURS: Eras
SPACE: Earth

Days and Seasons

The seasons are changes in the Earth's weather through the year. They are repeated at about the same time every year because the Earth takes one year to travel around the Sun. This causes the same part of the Earth to face the Sun at the same time each year. The Earth also spins all the way around on its axis every 24 hours, making day and night happen.

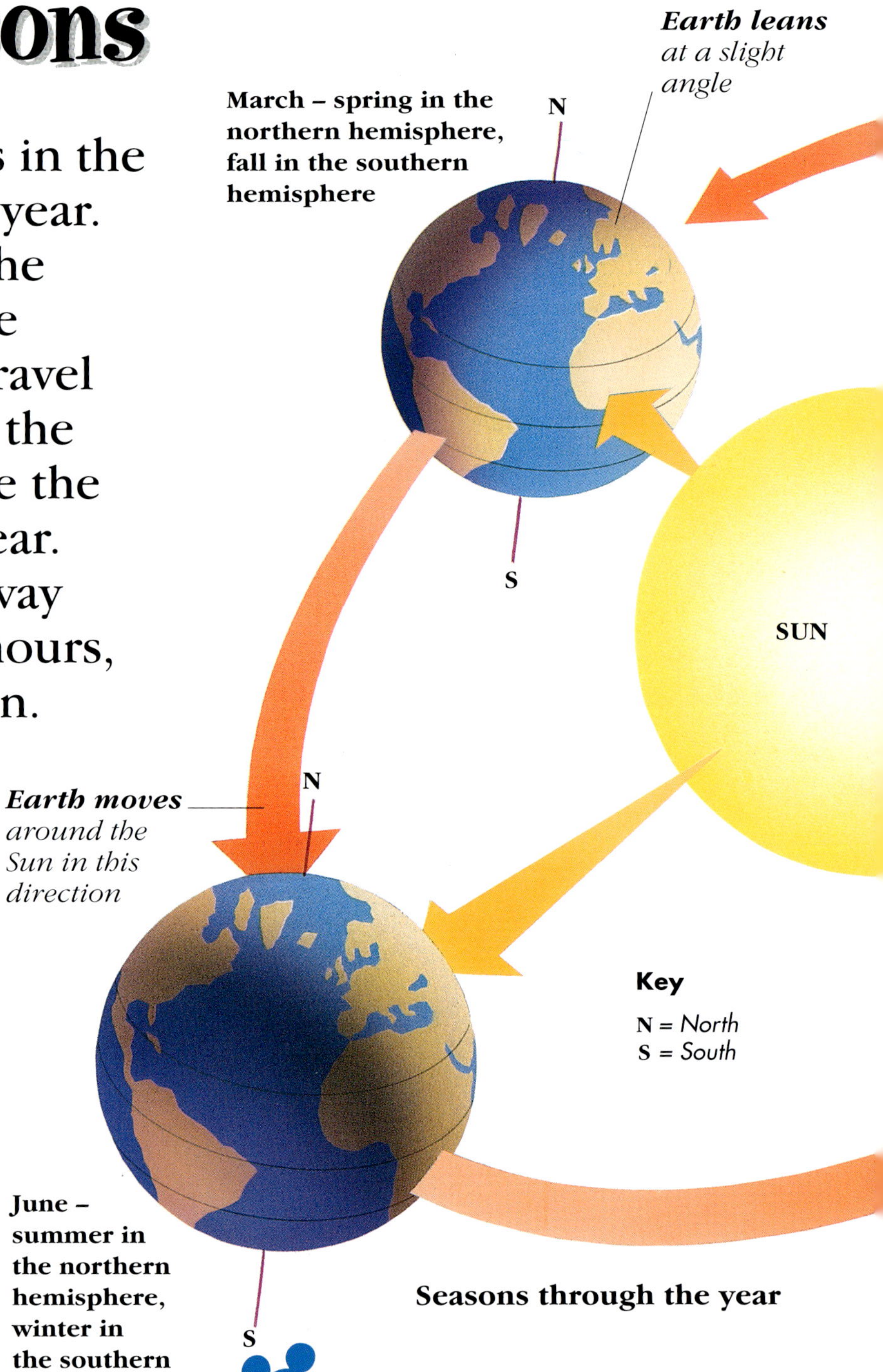

Seasons through the year

CHANGING SEASONS

The way the Earth leans at an angle means that at different times of the year some places are nearer the Sun than others. Places nearer the Sun get more light and heat, which creates their summer season. Places leaning away from the Sun have their winter season.

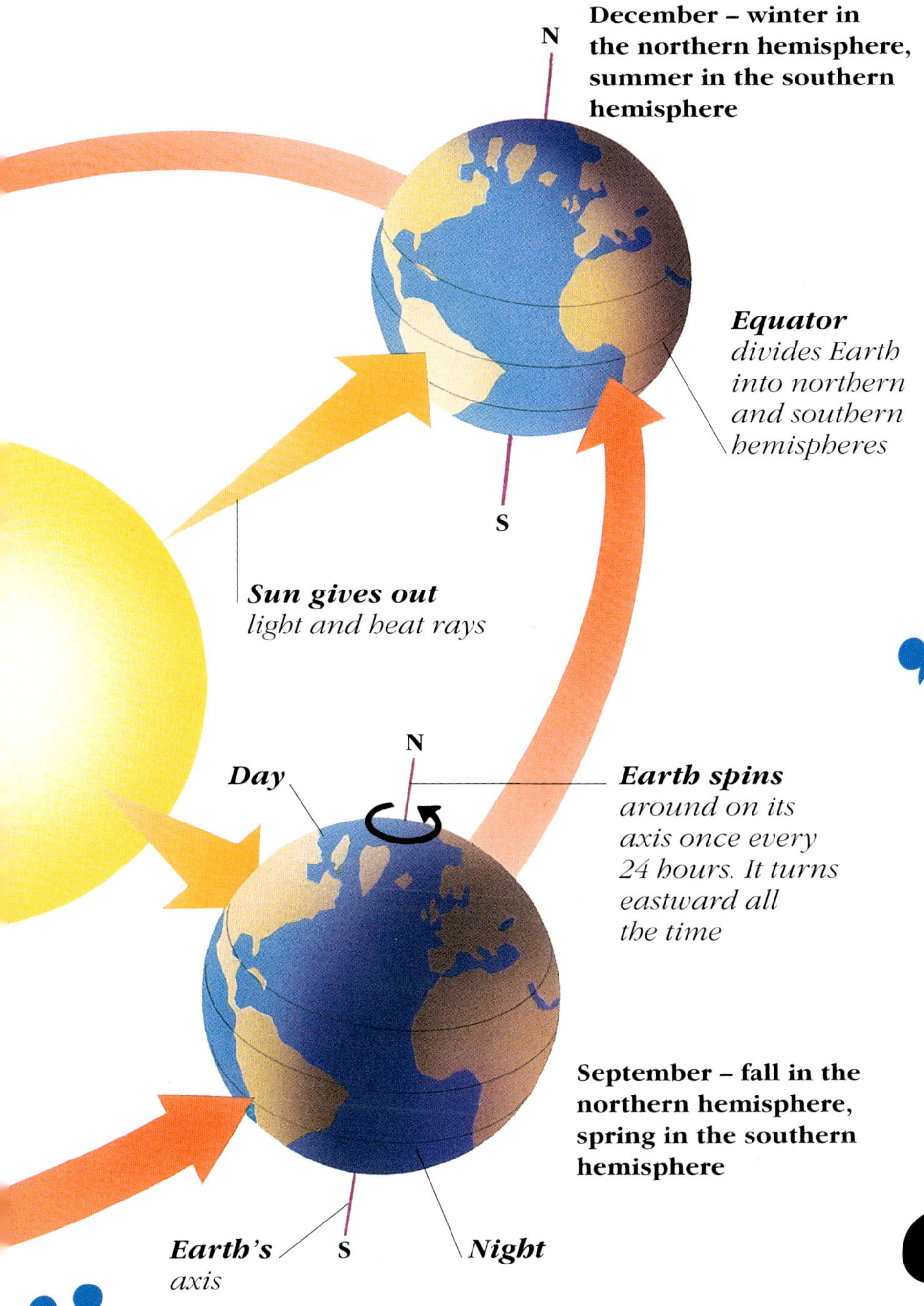

December – winter in North America in the northern hemisphere

December – summer in Australia in the southern hemisphere

WORLD SEASONS

The Earth is divided into two halves, called hemispheres, by the Equator. Most parts of the world have four seasons – spring, summer, fall, and winter. However, the two poles, at the far north and south of the planet, have only winter and summer. Places in the northern hemisphere have opposite seasons to places in the southern hemisphere.

DAY AND NIGHT

It takes 24 hours for the Earth to spin around once on its axis. The axis is an imaginary line through the center of the planet, joining the North and South Poles. The part of the Earth that faces the Sun has daytime. As the Earth spins, that part moves out of the light and into the dark. When it is day on one side of the Earth, it is night on the other.

Sunrise: a new day

AMAZING FACTS

★ **The Earth travels at more than 105,000 km/h (65,000 mph) as it whizzes around the Sun.**

FIND OUT MORE
SPACE: Sun
SCIENCE ALL AROUND US: Shadows

Inside and Out

The Earth is made up of layers of very hot rock and metal. Around the outside of the Earth are layers of gases. These gases are called the Earth's atmosphere. The atmosphere controls the temperature of the Earth and provides air for living things to breathe. It also keeps out some of the Sun's rays that can harm life on the planet.

Layers of Rock and Metal

The Earth is made up of three main layers. The thin outer crust we live on is made of cool, solid rock. Underneath the crust is a very thick, hot layer called the mantle. In the middle of the Earth is the core. It is made up of a solid metal inner core surrounded by a hot, liquid metal outer core.

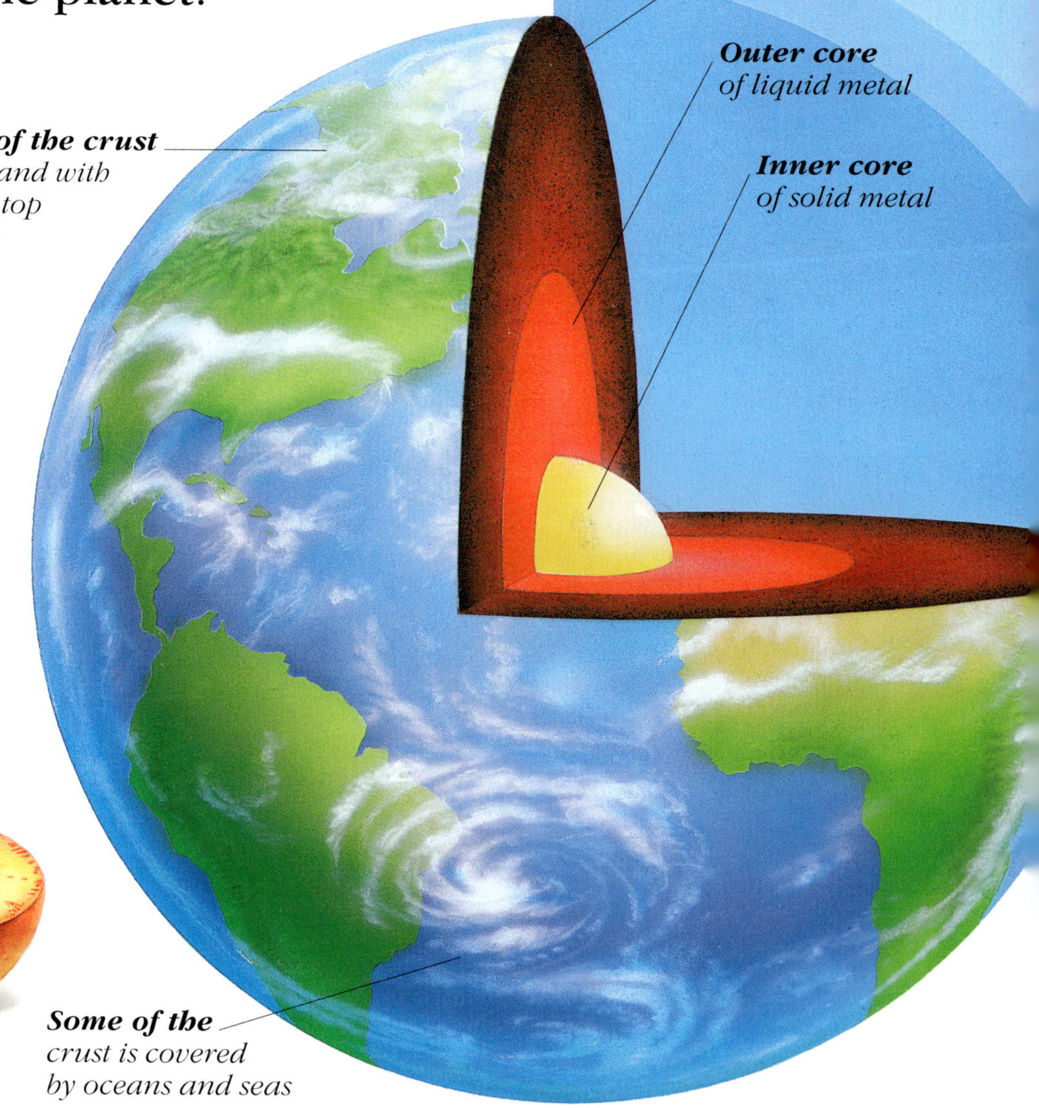

Earth's layers and atmosphere

The Earth Is a Peach

The layers of the Earth are like those of a peach. The thin skin of a peach is like the Earth's crust. The soft peach flesh is like the Earth's mantle. And the hard peach stone is like the Earth's hard metal core.

Layers of a peach

LAYERS OF GASES

There are five main layers of gases in the Earth's atmosphere. They are the troposphere, stratosphere, mesosphere, thermosphere, and exosphere.

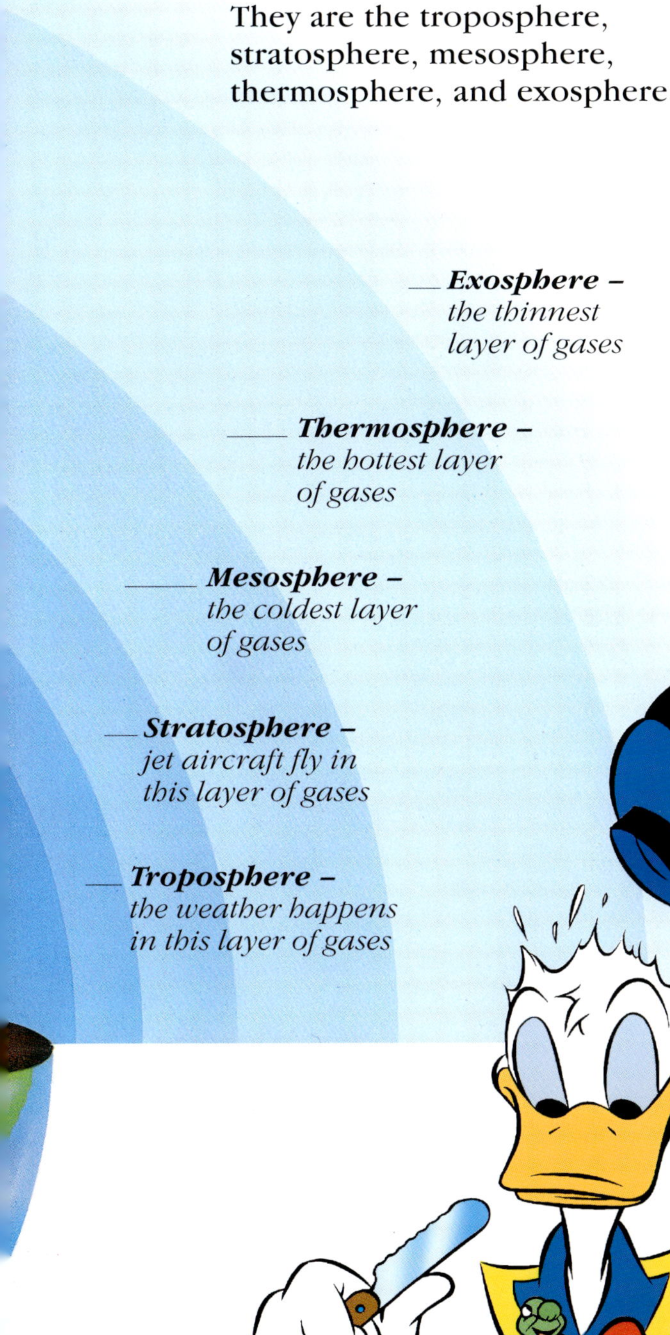

Gravity is the force that pulls things down to Earth

GRAVITY

When you bounce on a trampoline, a force called gravity pulls you back down. In the same way, Earth's gravity pulls everything toward its center. It stops things from floating away into Space and makes rain fall to the ground.

AMAZING FACTS

★ **If a person could walk to the center of the Earth, it would take them 10 weeks without stopping.**

FIND OUT MORE
SCIENCE ALL AROUND US: Gravity
SPACE: Earth's atmosphere

Moving Continents

The Earth's crust is cracked into giant pieces of rock called plates. The plates fit together like the pieces of a jigsaw puzzle. When the liquid rock deep inside the Earth moves, it makes the plates slide about. On top of the plates are huge areas of land called continents. As the plates slide, the continents move, too.

SLIDING PLATES

When two plates slide past each other, a huge crack called a fault appears in the rock. There is a long crack on the western coast of North America, called the San Andreas fault. Earthquakes often happen along fault lines.

San Andreas fault

CHANGING EARTH

More than 200 million years ago, all the land on Earth was joined together in one huge mass. Scientists call it Pangaea. Slowly, it split apart and the pieces drifted across the planet to make the seven continents of the world today. The continents still move a little every year.

How Pangaea split up to make today's continents

Slowly, the continents drifted across the Earth

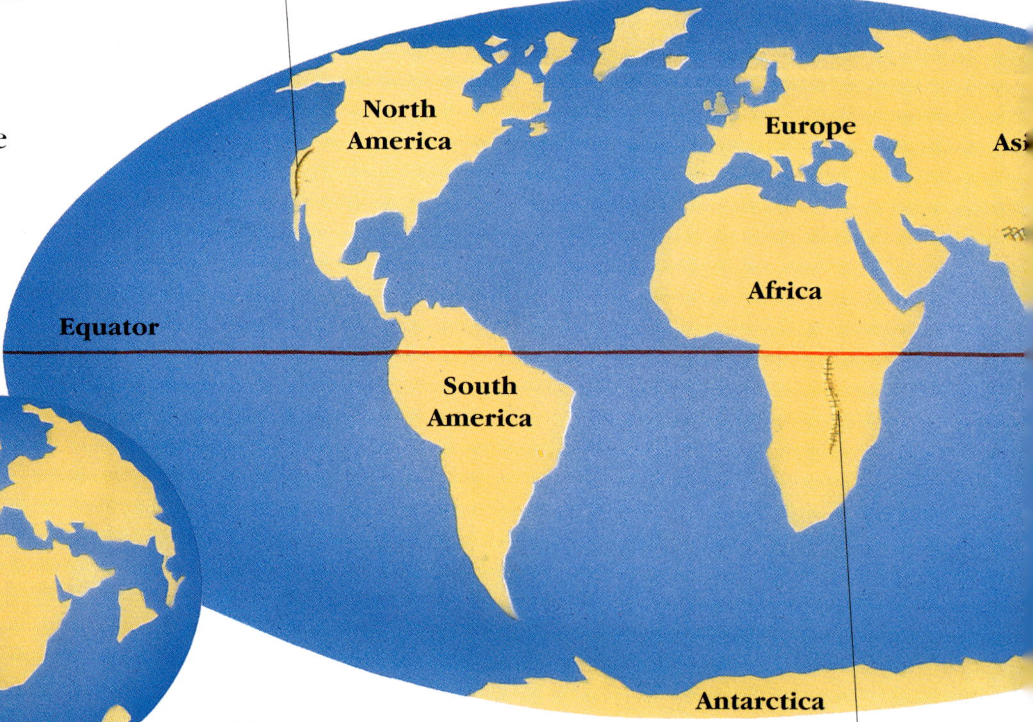

World's continents as they are today

The San Andreas fault *lies where two plates slide past each other in California, U.S.. The fault line is about 1,125 km (700 miles) long*

East African Rift Valley *was made when two plates were pulled apart*

Plate Movement

When two of the Earth's plates touch, one of three things may happen – they may crash into each other, they may pull apart, or they may slide past each other.

Most plates move apart under the oceans, and liquid rock rises up to fill the gap

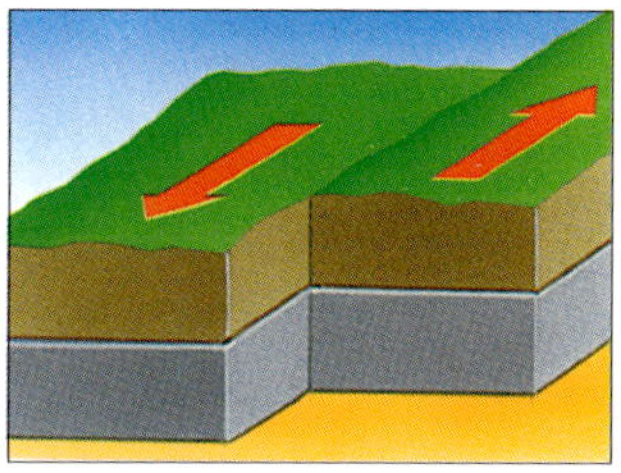

Faults form and earthquakes may happen when plates slide past each other

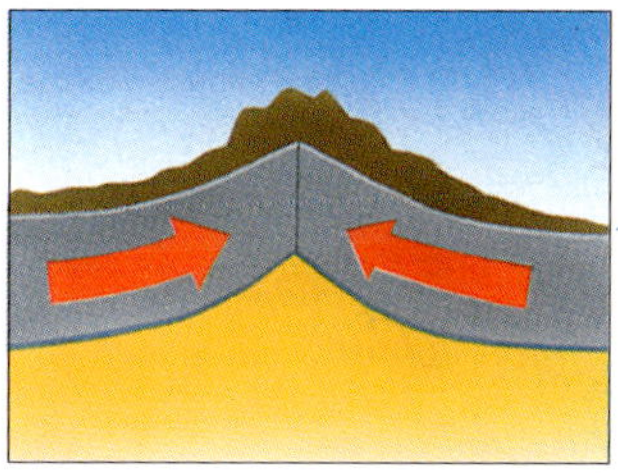

Mountains may be forced up when plates crash together

The Marianas Trench *is the deepest point on Earth. It formed when two plates crashed and one was pushed down inside the Earth*

The Himalayas *were created when two plates crashed, making their edges crumple up*

Only in Australia

Australia became a separate continent about 45 million years ago. Ordinary mammals, such as elephants and bears, replaced marsupials, or mammals with pouches, in most other parts of the world. But they did not develop in Australia. Instead its ancient marsupials evolved to form new species, such as koalas and kangaroos, which are now found only there.

Amazing Facts

★ Every year, the world's continents move about 2 cm (¾ in).

FIND OUT MORE
FAMOUS PLACES: Himalayas
MAMMALS: Marsupials

Earthquakes

An earthquake is a sudden shaking of the Earth's crust. Most earthquakes happen when the edges of plates slide past each other. The rocks jerk suddenly or shudder, making the surface of the Earth shake. There are about half a million earthquakes each year, but only about a thousand cause any serious damage. When earthquakes happen under the sea, they sometimes make enormous waves, called tsunami.

During an Earthquake

The shock waves from an earthquake spread out from a point underground called the focus. The worst damage happens at a point on the Earth's surface called the epicenter. It lies above the focus, often on a fault line.

Earthquake damage in Kobe, Japan, 1995

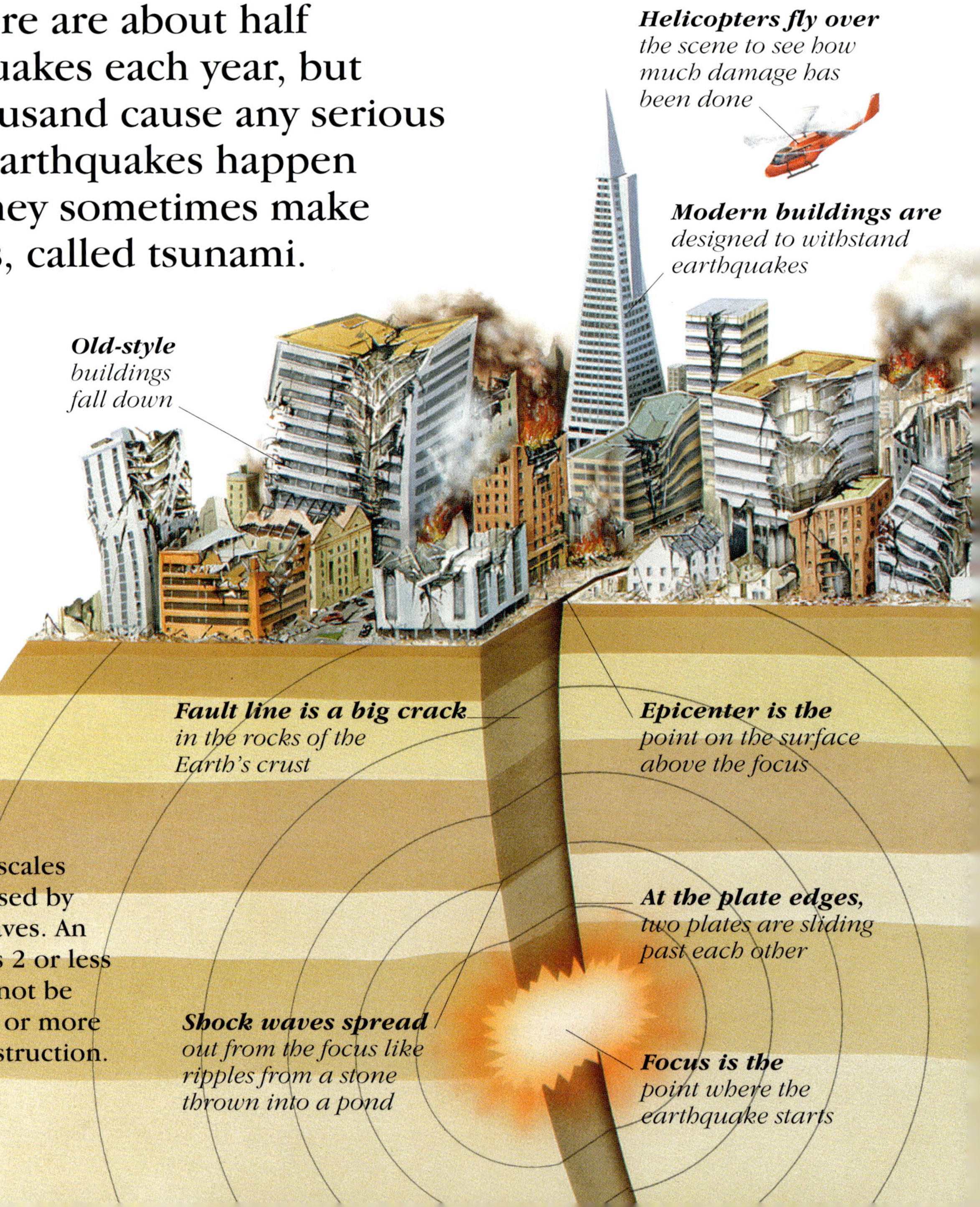

Earthquake Energy

The Richter and Mercalli scales measure the energy released by an earthquake's shock waves. An earthquake that measures 2 or less on the Richter scale may not be felt. One that measures 7 or more can cause widespread destruction.

EARTHQUAKE STRENGTH

The Mercalli scale measures the strength of an earthquake by looking at the damage caused. There are 12 levels of damage on the scale.

Level 3 **Slightly felt –** ceiling lights shake

Level 5 **Felt by everyone –** liquid spills

Level 9 **Severe damage –** towers fall, pipes break

Level 12 **Everything damaged –** big areas of land slip and move

Four of the 12 levels of the Mercalli scale

Fires are common *after an earthquake*

Imaginary earthquake in San Francisco, U.S.

Transamerica Building in San Francisco, U.S.

SAFE FROM EARTHQUAKES

Some buildings are specially designed to stay standing during an earthquake. They have strong foundations and can move a little if the Earth shakes. The pyramid shape of the Transamerica Building in San Francisco, U.S., helps to stop it from toppling over in an earthquake.

AMAZING FACTS

★ **Some animals behave strangely before an earthquake. Dogs bark wildly because they may be able to sense the shaking of the ground before people can.**

FIND OUT MORE
CHILDREN OF THE WORLD: Japan
GREAT INVENTIONS: Skyscrapers

Volcanoes

A volcano is a hole in the Earth's crust through which hot, liquid rock pours out from inside the planet. As the rock cools down, it makes hills and mountains. Most volcanoes are found on the edges of the Earth's plates. From time to time an active volcano erupts, or explodes. A volcano that has not erupted for many years is said to be extinct.

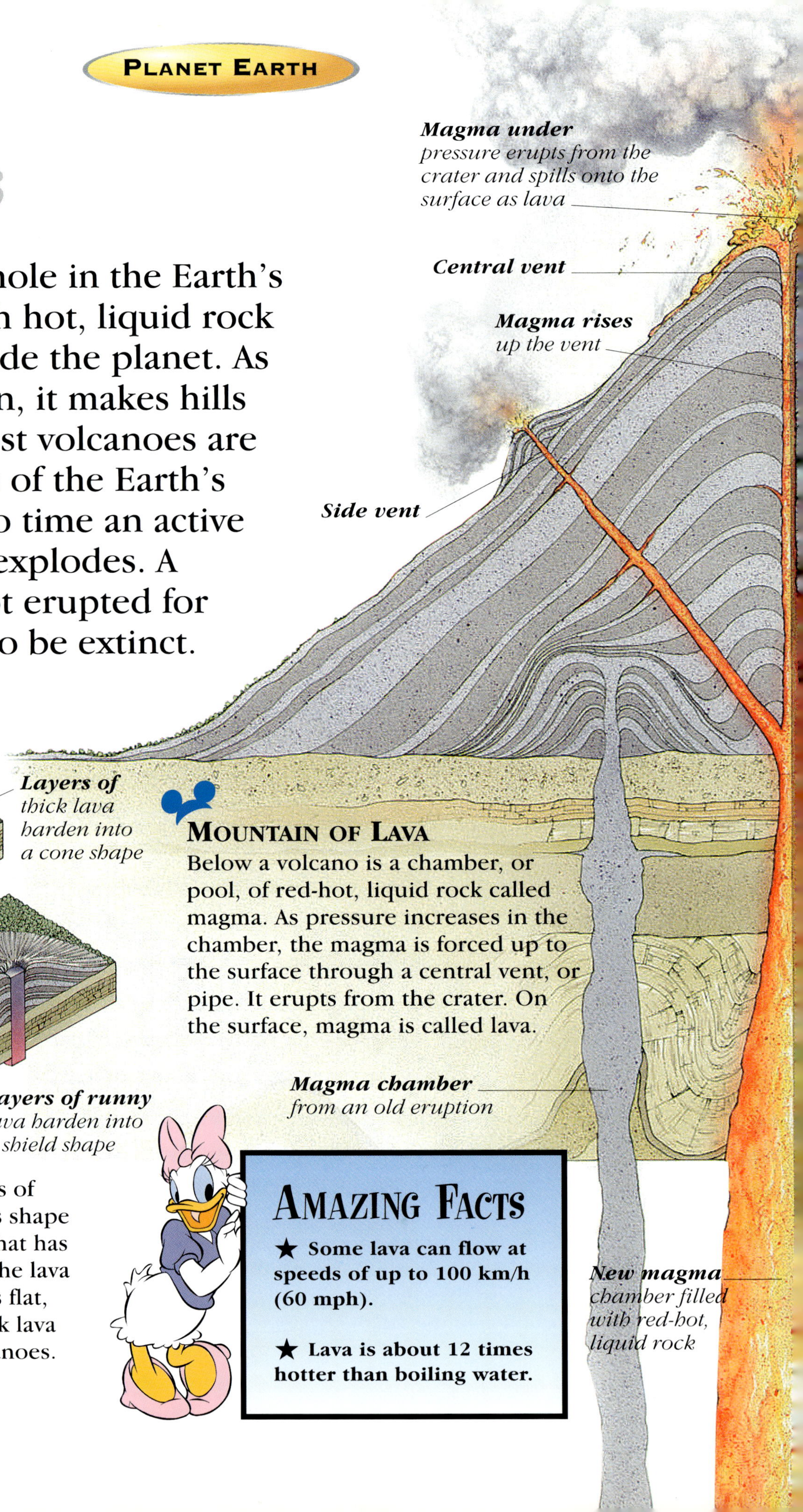

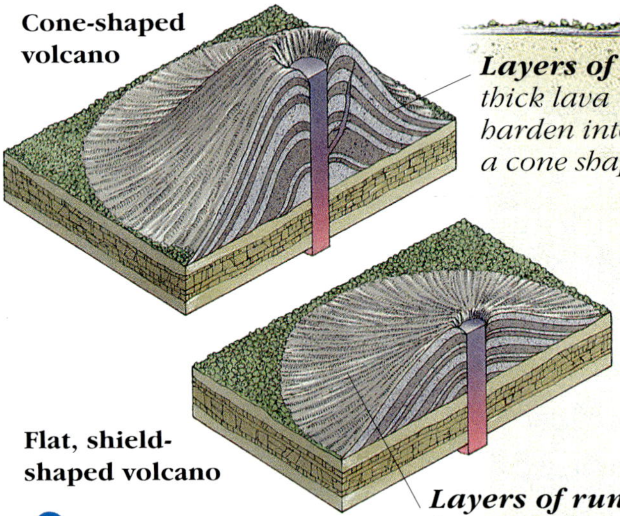

MOUNTAIN OF LAVA

Below a volcano is a chamber, or pool, of red-hot, liquid rock called magma. As pressure increases in the chamber, the magma is forced up to the surface through a central vent, or pipe. It erupts from the crater. On the surface, magma is called lava.

VOLCANO SHAPES

A volcano is made up of layers of old lava that have set hard. Its shape depends on the type of lava that has poured from it, and how far the lava has flowed. Runny lava makes flat, shield-shaped volcanoes; thick lava makes tall, cone-shaped volcanoes.

AMAZING FACTS

★ **Some lava can flow at speeds of up to 100 km/h (60 mph).**

★ **Lava is about 12 times hotter than boiling water.**

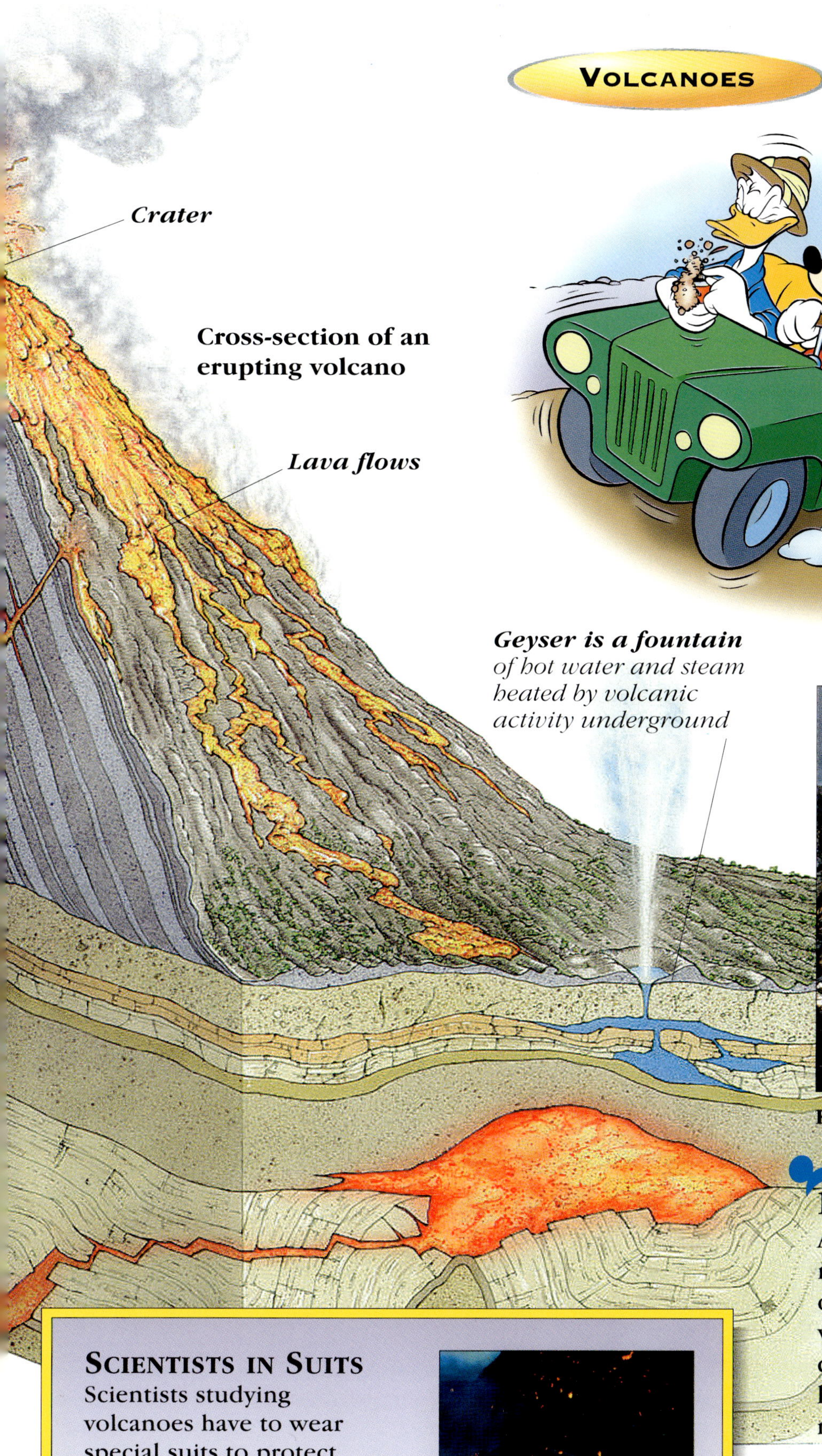

Kelimutu Crater Lakes, Indonesia

BLOWING ITS TOP

A fierce eruption can blow the top right off a volcano. This leaves a crater, or basin, called a caldera. It can fill with water to make a crater lake. If the crater stays dry, small cones of ash and lava can build up on its flat floor from more volcanic eruptions.

SCIENTISTS IN SUITS

Scientists studying volcanoes have to wear special suits to protect them from the red-hot lava. Samples of the lava help them to find out more about rocks inside the Earth and to predict new eruptions.

Scientist studying lava flow

FIND OUT MORE
CHILDREN OF THE WORLD: Geysers
STORY OF THE PAST: Pompeii

Mountains

Most mountains are formed when the Earth's plates move. There are three main types of mountains: fold mountains, block mountains, and dome mountains. Many of the highest mountains, such as the Himalayas in Asia, are fold mountains. They lie in long lines, called ranges. Some mountains are still growing, but others are being worn away by the weather.

How the Himalayas Formed

1 About 45 million years ago, the plate carrying India was moving toward the plate carrying the rest of Asia.

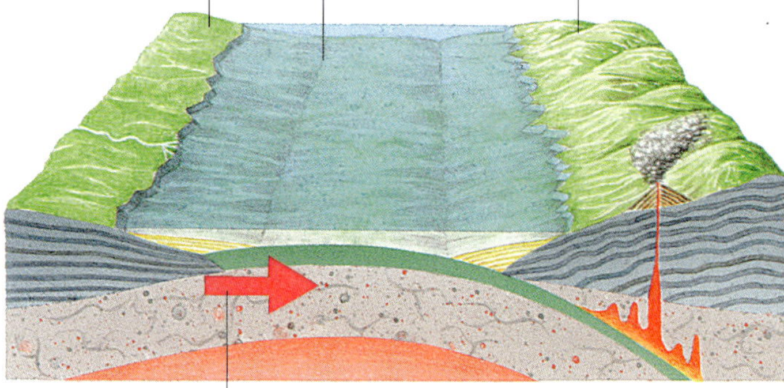

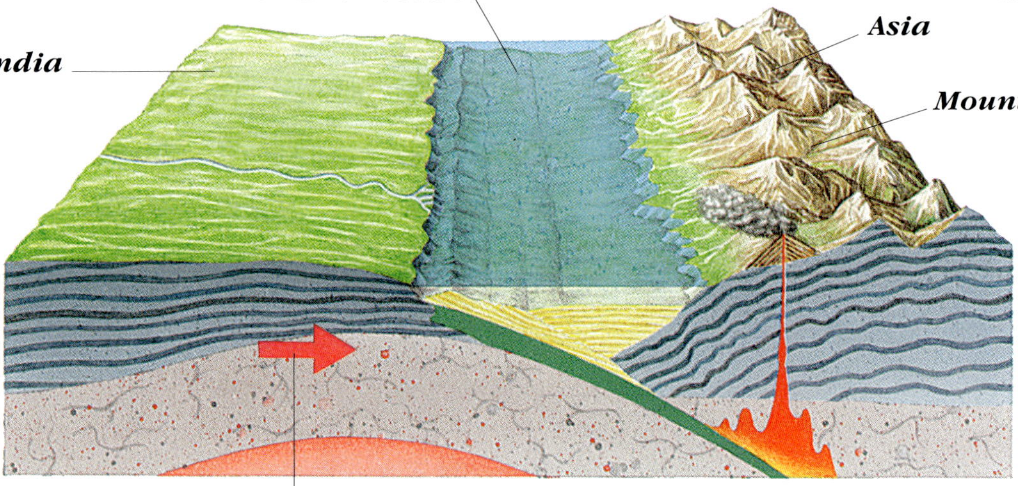

2 About 40 million years ago, the Indian plate crashed into the Asian plate. The rocks on the ocean floor were squeezed together and crumpled up into giant folds.

3 Eventually, India joined onto Asia, and the ocean disappeared. A huge range of fold mountains, the Himalayas, now marks the place where the two plates crashed together.

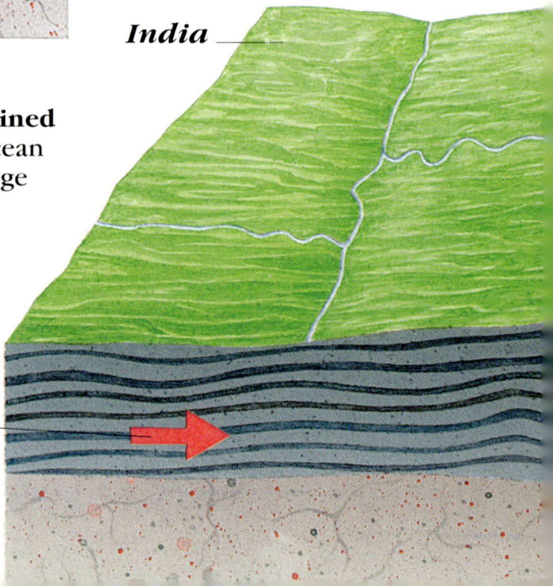

Amazing Facts

★ **The Himalayas are still rising by about 1 cm (½ in) every year.**

Block mountain formation

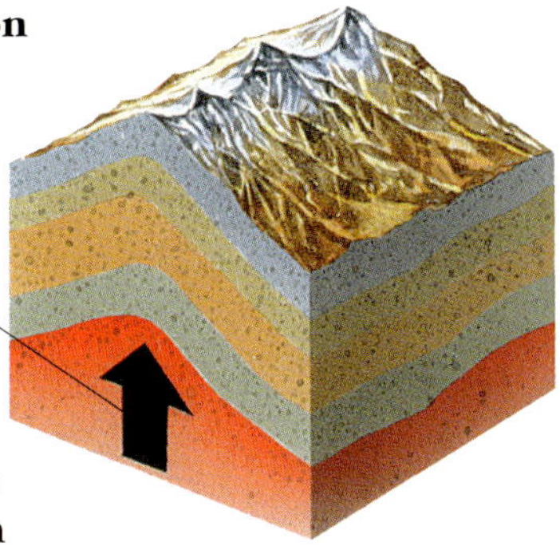

Dome mountain formation

MOUNTAIN AIR

Earth's atmosphere is made of air, a mixture of nitrogen, oxygen, and other gases. High up on mountains, the air is "thin" – its gases are more spread out than at ground level, so it contains less oxygen for breathing.

Peruvians are used to the thin air in the Andes Mountains

OTHER TYPES OF MOUNTAINS

Block mountains are formed when a block of rocks is forced up between two fault lines. Dome mountains form when magma pushes surface rocks above it upward into a bulge.

Making a model of a dome mountain out of modeling clay

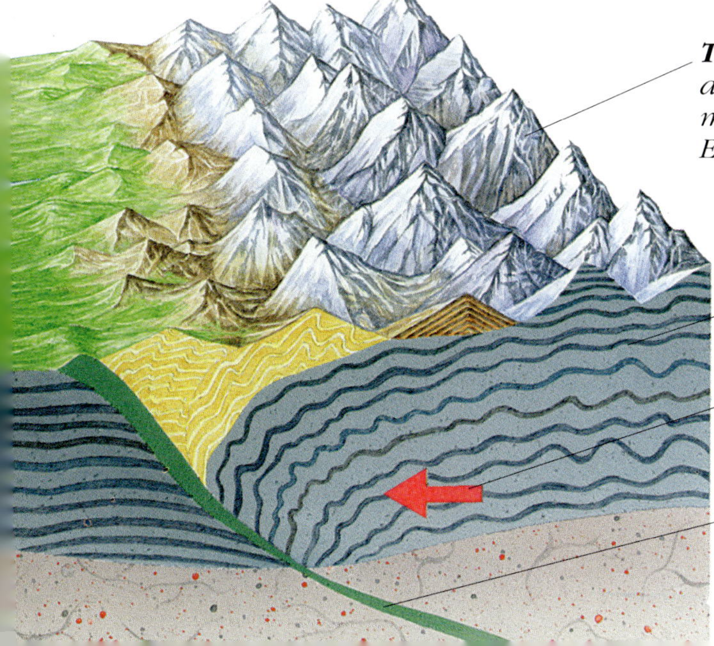

FIND OUT MORE
FAMOUS PLACES: Mountains
PLANT LIFE: Mountain plants

Rocks and Fossils

Rocks of the Earth's crust are being made and worn away all the time, in a slow process called the rock cycle. When magma comes to the surface, it hardens into igneous rocks. These are eroded, or worn away, into tiny particles that are washed off the land into rivers, then carried along as sediment toward the sea. Where layers of sediment build up and harden, sedimentary rocks form.

Most rocks are made from mixtures of particles called minerals. Rocks form because these particles stick together. Minerals can be recognized by their special colors, hardness, and the kinds of crystals they form.

Many rocks contain crystals of quartz, the most common mineral

Schist is a metamorphic rock formed when sedimentary clays are squeezed together

COOLED ROCKS

Igneous rocks are formed when magma, or molten rock, from inside the Earth cools and hardens. Some igneous rocks are formed under the ground, others form on the Earth's surface from lava.

Gabbro is an igneous rock formed when magma cools

CHANGED ROCKS

When igneous or sedimentary rocks beneath the Earth's surface are squeezed under great pressure, or heated to a high temperature, they change into metamorphic rocks. Rocks in volcanic areas are often changed into metamorphic rocks.

HOW COAL IS FORMED

The fossil fuel coal began to form about 300 million years ago. It is made from dead plants that were buried and slowly squashed into hard rock.

1 Swamp plants died and sank into water

2 They were buried in mud and pressed down, forming peat

3 The peat was buried deeper and pressed to form soft brown coal

4 Finally, hard, shiny black coal formed

Fossil of an ammonite, a sea creature

FORMING FOSSILS

Fossils are the remains of plants and animals preserved mainly in sedimentary rocks. For a fossil to form, the plant or animal must be buried quickly by sediment. Over millions of years, the sediment hardens into rock. The plant or animal rots, leaving an impression in the rock, or becomes fossilized into stone itself.

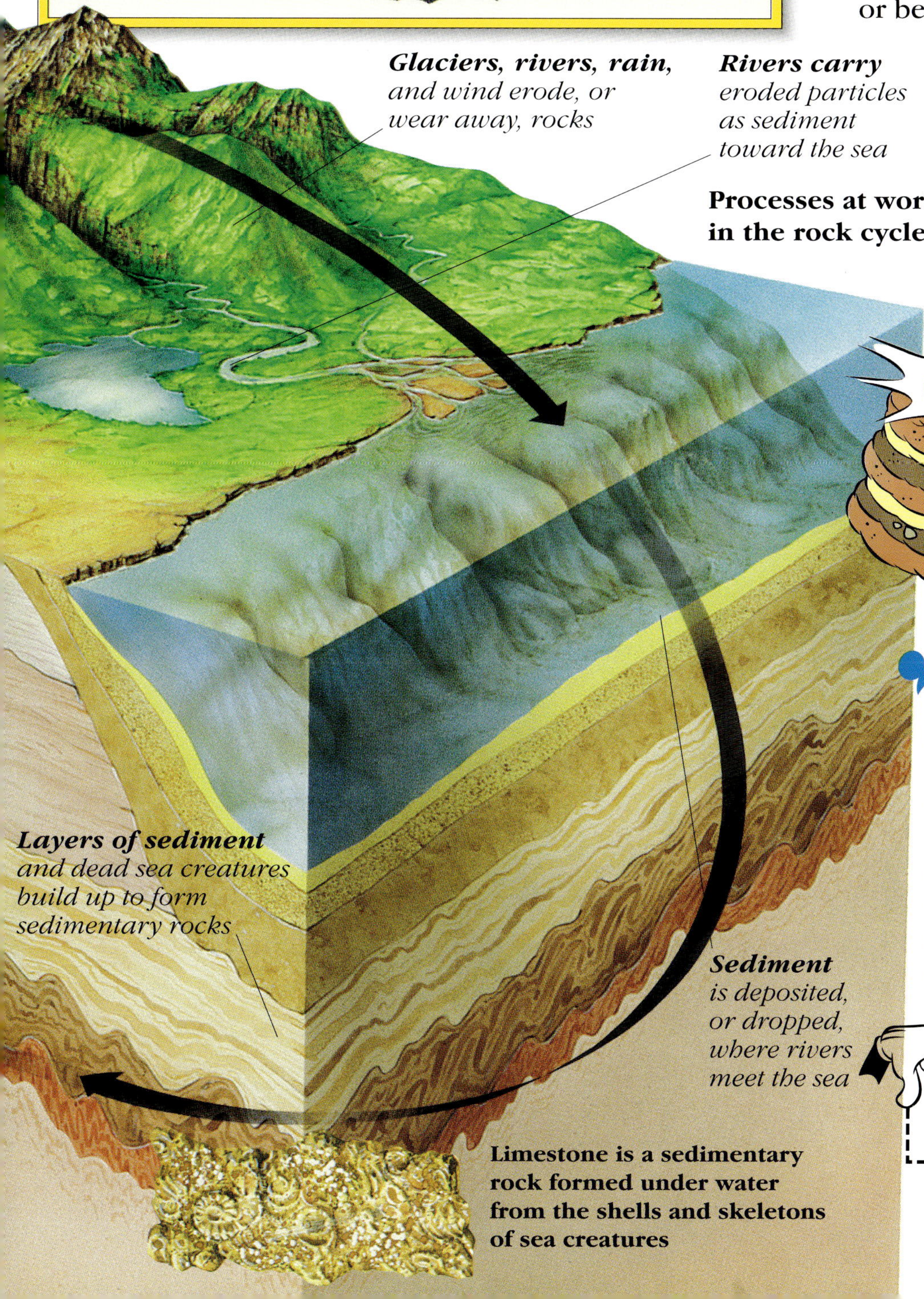

Processes at work in the rock cycle

LAYERED ROCKS

Sedimentary rocks are formed by the build-up of layers, or strata, of sediment that are squashed into hard rock. Some sediment comes from rocks that have been worn down by erosion. Other sediment comes from shell and coral fragments in the sea.

FIND OUT MORE
DINOSAURS: Fossils
STORY OF THE PAST: Mines

Glaciers

Glaciers are huge rivers of ice that form from snow that falls on high ground. They creep slowly downhill, wearing away rocks as they move. When the climate warms up or when the snout, or end, of a glacier flows into a warmer area, the glacier begins to melt. As it melts, it drops the worn-away rock it is carrying.

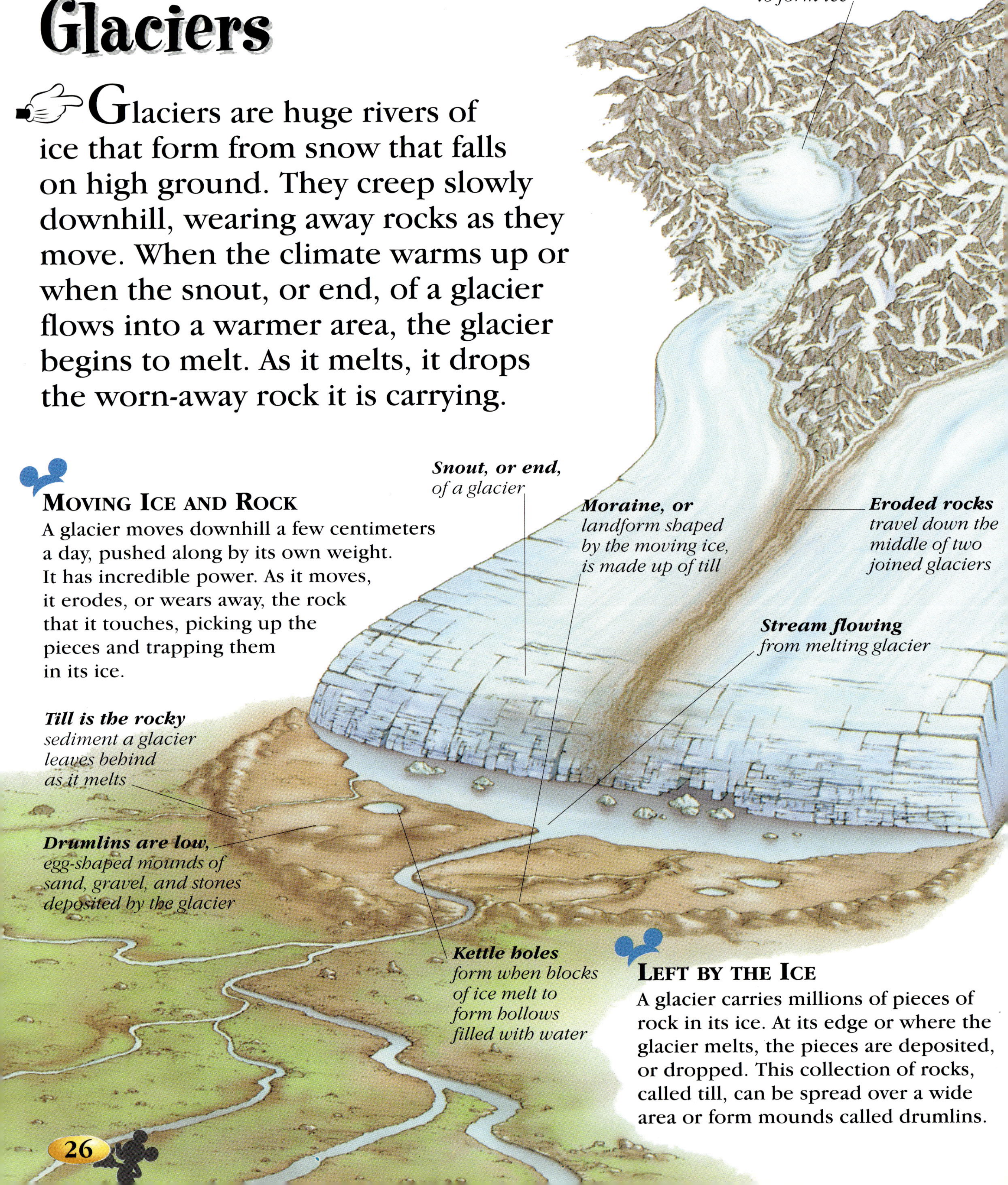

MOVING ICE AND ROCK

A glacier moves downhill a few centimeters a day, pushed along by its own weight. It has incredible power. As it moves, it erodes, or wears away, the rock that it touches, picking up the pieces and trapping them in its ice.

LEFT BY THE ICE

A glacier carries millions of pieces of rock in its ice. At its edge or where the glacier melts, the pieces are deposited, or dropped. This collection of rocks, called till, can be spread over a wide area or form mounds called drumlins.

SHAPING VALLEYS

1 **In warmer times before the glacier** formed, a river flowed downhill. It wore away a valley with sides shaped like the letter "V."

2 **The climate became colder. Ice** formed and filled the valley with a glacier. The glacier pushed downhill, wearing away more of the valley.

3 **When the climate became warmer** again, the glacier melted. It left behind a deeper, straighter valley with a "U" shape and a flat floor.

AMAZING FACTS

★ **The longest glacier is the Lambert-Fisher Glacier in Antarctica. It is 515 km (320 miles) long.**

FIND OUT MORE
ATLAS OF THE WORLD: Antarctica
SCIENCE ALL AROUND US: Ice

Oceans and Seas

More than two-thirds of the Earth's surface is covered by the water of the oceans and seas. There are five great oceans – the Pacific, Atlantic, Indian, Southern, and Arctic – and many seas, which are smaller. The oceans and seas are always on the move. The wind blows and makes waves, and the Moon and Sun pull on the Earth to make the tides.

UNDER THE WATER

The ocean floor is not completely flat. Just as the Earth's dry land surface varies in shape, the ocean floor has mountains, ridges, and trenches of its own. Many of these features are created by plate movements and volcanic activity inside the Earth, just as they are on land.

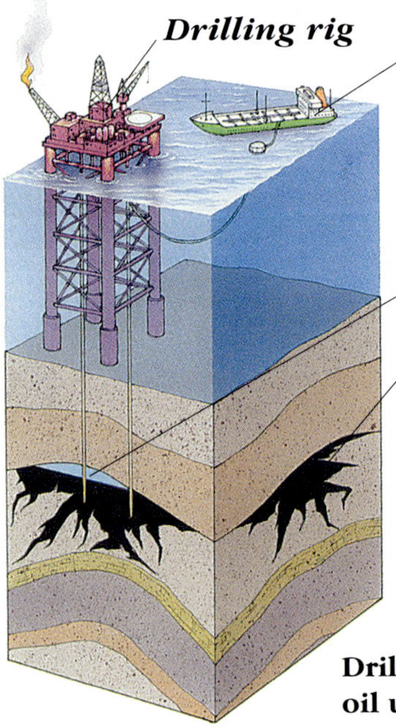

Drilling for gas and oil under the sea

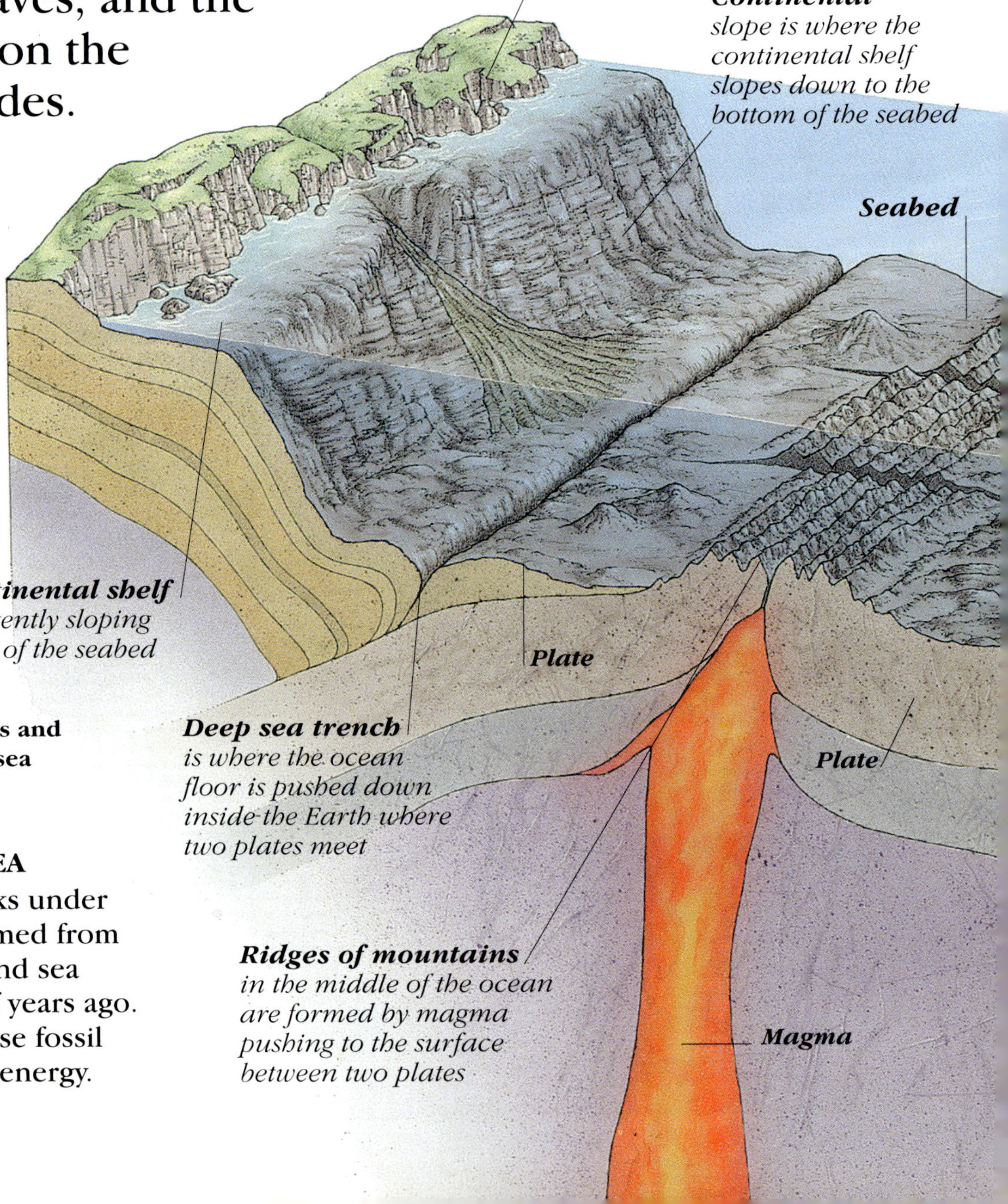

FUEL FROM UNDER THE SEA

Oil and gas are trapped in rocks under the sea and land. They are formed from the buried remains of plants and sea creatures that lived millions of years ago. People drill down to reach these fossil fuels, which are used to make energy.

THE TIDES

The water level of the oceans rises to make high tides and falls to make low tides. The tides are caused mainly by the Moon pulling on the oceans as it moves around the Earth. When the Earth, Moon, and Sun are lined up, the Moon and Sun both pull on the oceans, making the tides even higher.

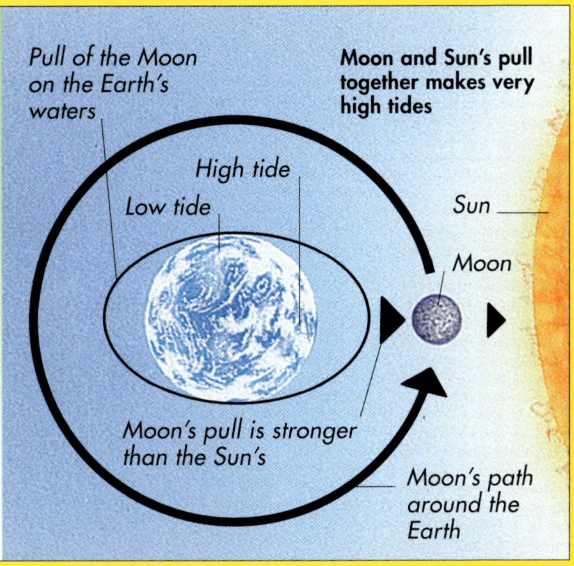

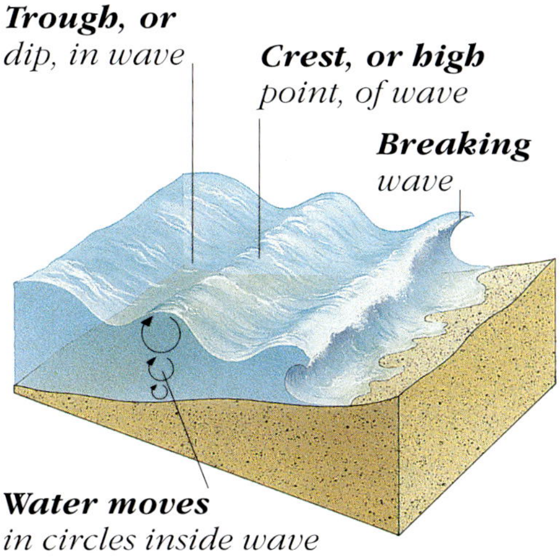

Moving waves

HOW WAVES FORM

Waves move across the surface of the sea. Their troughs and crests are made by the force of the wind as it blows over the sea's surface. When a wave reaches the shore, it breaks, or topples over, onto the beach.

Landscape of the ocean floor

AMAZING FACTS

★ At almost 11 km (6¾ miles) deep, the Marianas Trench, at the bottom of the Pacific Ocean, is 29 times the height of the Empire State Building in the U.S..

FIND OUT MORE
ANIMALS OF THE SEA: Shore birds
SPACE: Moon

Rivers and Lakes

A river is water that flows downhill along a narrow route called a channel. It is kept in its channel by banks of stone and soil, which act like walls. Water can have a lot of power to change the shape of the land. Rivers carve out valleys and make lakes and waterfalls. They wash eroded rock down from the mountains, dropping some of it on the land, before flowing to the sea.

Boats on the Ganges Delta, India

RIVER DELTAS

As a river nears the sea, it slows down and deposits sediment, such as sand, silt, and clay. This material, called alluvium, spreads out to make an area of flat, wet land called a delta.

FROM MOUNTAINS TO SEA

The source, or start, of a river is usually in the mountains, where rainfall is highest. Water always flows downhill, following the quickest route to the sea. At first a river cuts a deep valley, but as it reaches flatter land, it starts to cut sideways. Its channel becomes wider and it meanders, or winds, across a wide valley.

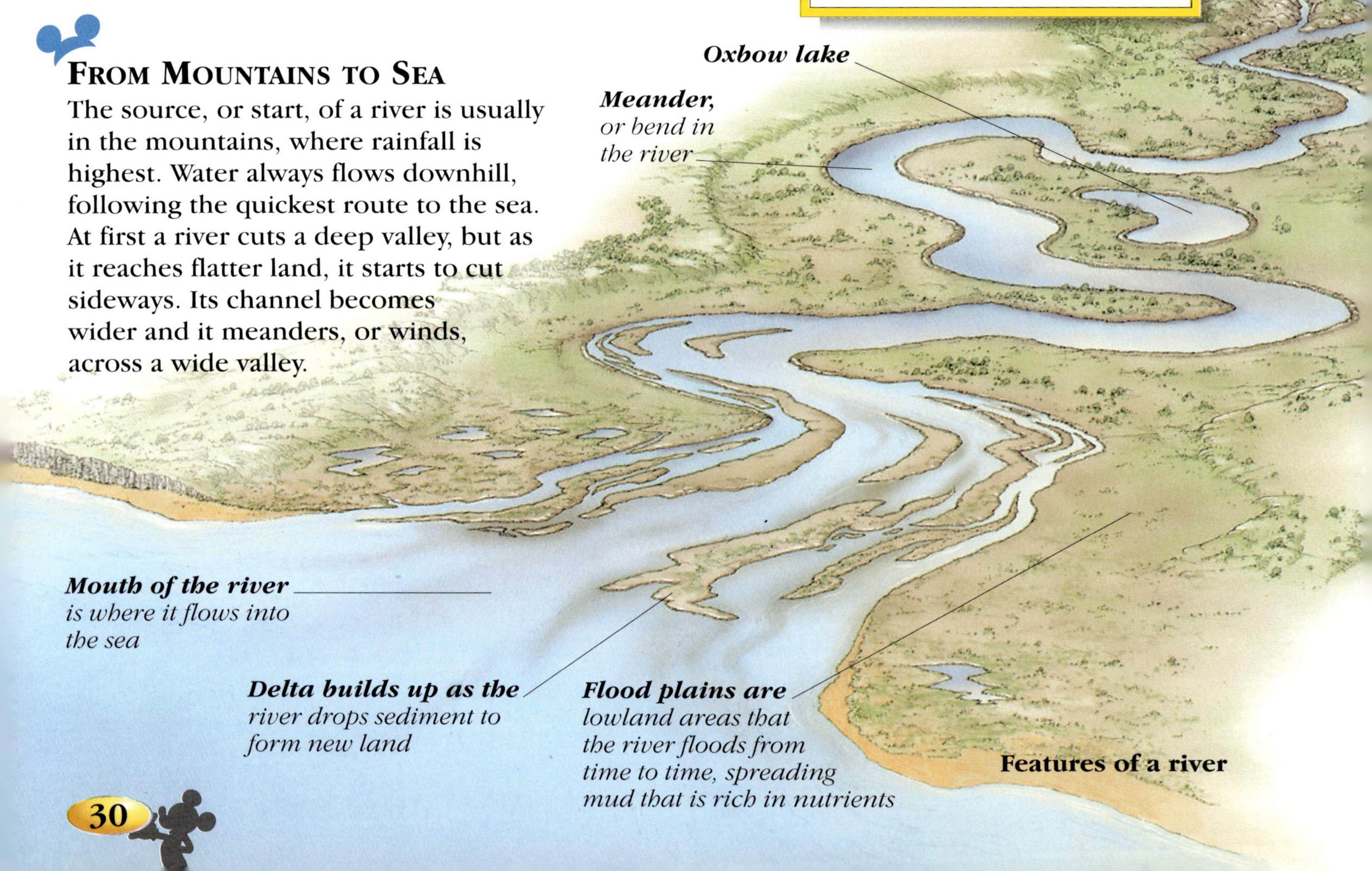

Features of a river

Amazing Facts

★ **The longest river is the Nile River, in Egypt. It is over 6,400 km (4,000 miles) long.**

How a Waterfall Makes a Gorge

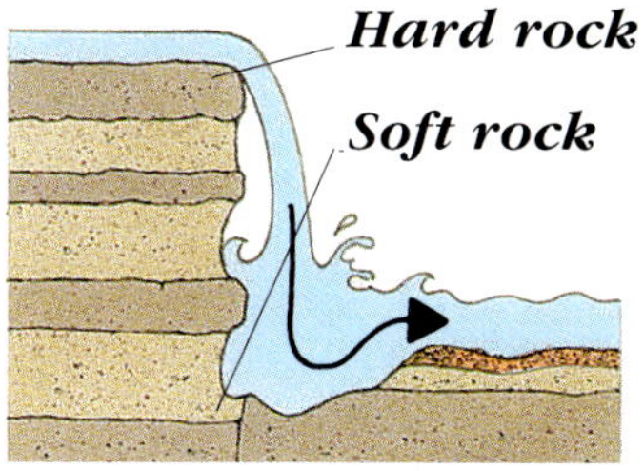

1 **A waterfall flows over** a layer of hard rock that will not wear away easily, and onto a layer of softer rock below.

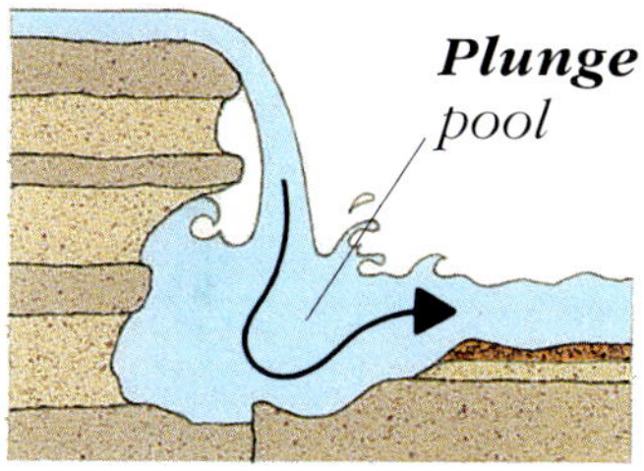

2 **As water swirls** around on the softer rock, it wears away a deep plunge pool.

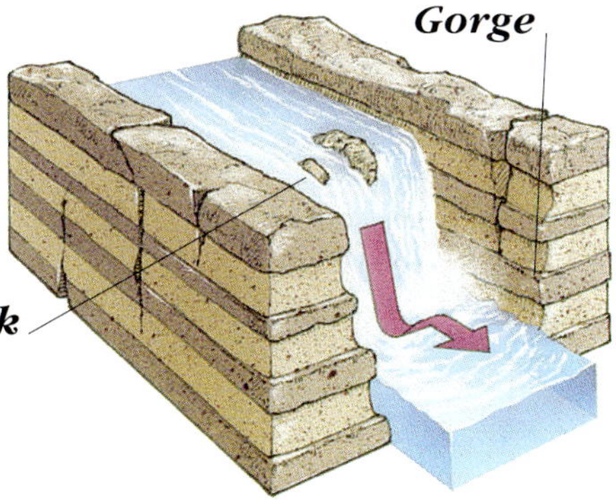

3 **Over a long time, the** river wears away some of the hard rock, too, and the waterfall moves back upstream, creating a gorge.

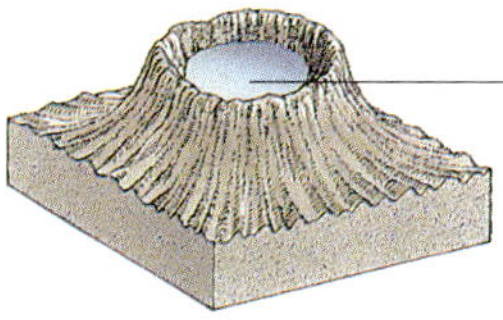

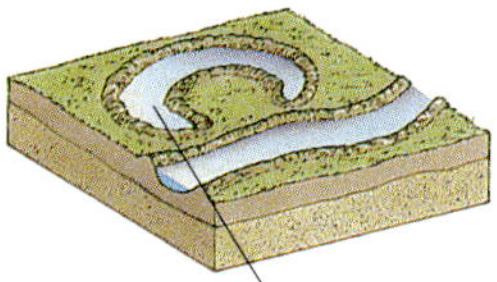

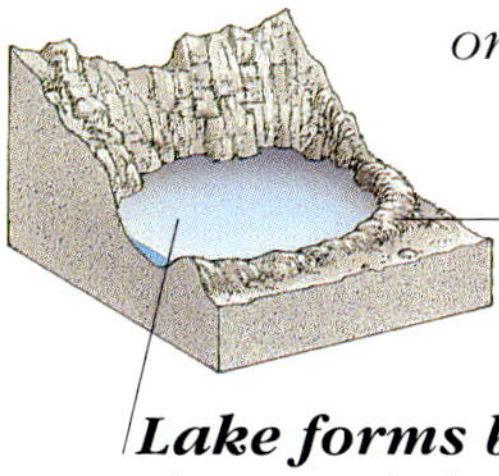

How Lakes Form

When water collects in a dip in the land, a lake is formed. This can happen in different ways. A volcano crater may fill with water; a river may cut through a bend, leaving behind a curved oxbow lake; or the mud and stones left behind by a glacier may form a dam that traps water behind it.

FIND OUT MORE
FAMOUS PLACES:
Waterfalls, Ganges River

The Changing Landscape

The shape of the land is always changing because water, ice, and wind wear it away. This process is called erosion or weathering. It usually happens very slowly, but soft rocks, such as limestone, wear away faster than hard rocks, such as granite. Rocks covered in plants have more protection from erosion.

The red limestone rocks of Bryce Canyon, U.S., have been worn into shapes that look like stone castles or temples. Thousands of years of rain, running water, freezing and thawing, and chemicals in the air have eroded the rocks into these strange shapes.

Limestone towers, Guilin, southwestern China

Rock formations in Bryce Canyon, U.S.

WATER SCULPTURE

Water can wear away rock by washing against it or dissolving it chemically. Chemical weathering can make dramatic landscapes, such as the limestone towers and caves at Guilin, China.

HOW LIMESTONE CAVES ARE FORMED

Caves can be formed in limestone by water sinking under the ground and wearing away the soft rock. Acids in rainwater dissolve, or chemically wear away, limestone. Inside the caves, dripping water forms limestone "icicles," called stalactites, and pillars, called stalagmites. Stalactites grow down from the roof and stalagmites grow up from the floor.

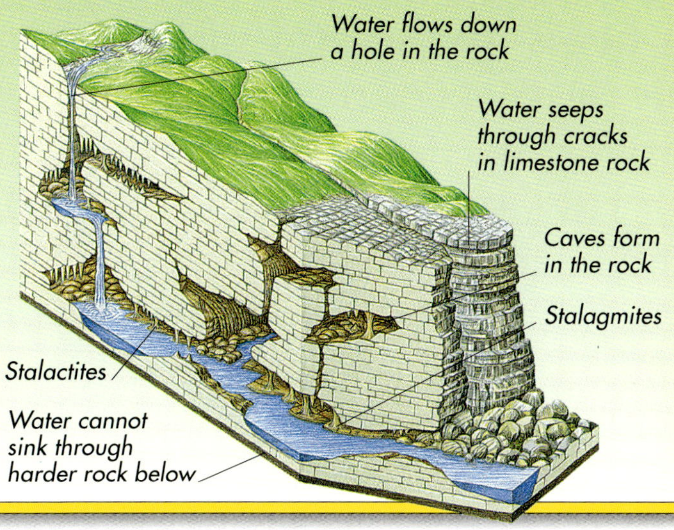

THE POWER OF THE WIND

Wind erosion is common in dry areas, such as deserts, because there are few plants to slow down the wind or cover up rocks. In deserts, the wind blows sharp grains of sand against the rocks, carving shapes such as columns and arches.

Sculpted rock caused by wind erosion

FIND OUT MORE
FAMOUS PLACES: Erosion
PLANT LIFE: Soil

What Makes the Weather?

The Sun makes the Earth's weather. It heats up the atmosphere, making the air move as currents or winds. It also drives the Earth's water cycle and the world's ocean currents. Together, the currents, winds, and water cycle keep heat and water moving around the Earth, making the world's climates, or weather patterns.

World's main bands of winds

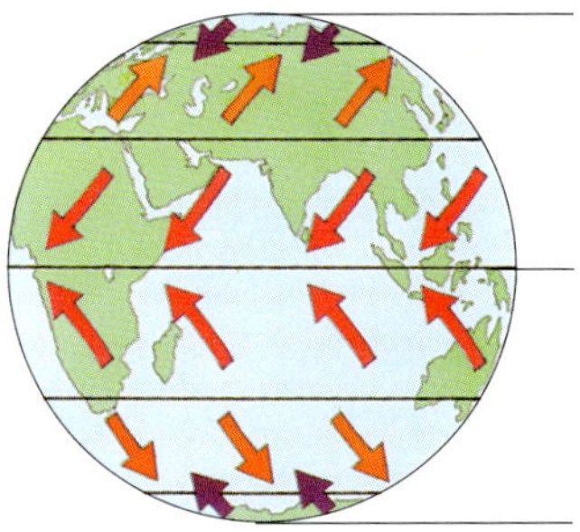

The World's Winds

Wind influences the weather. There are three main bands of winds around the Earth: trade winds blow toward the Equator; easterlies blow from the east; and westerlies blow from the west.

Water Everywhere

The water cycle is the movement of water from the Earth, up into the sky, and back down again. It works when the Sun heats water on land or in the sea, making the water turn into water vapor – an invisible gas in the air. High in the sky, the water vapor cools and turns into water droplets and ice crystals that collect into clouds.

Water cycle

THE EFFECTS OF EL NIÑO

Moving currents of water in the oceans affect the world's weather. For example, a warm current called El Niño appears off the northwestern coast of South America every few years. Warm water builds up near the coast instead of moving westward toward Australia and Indonesia. This can change the amount of heat and moisture in the air around the world, causing floods in the Americas and drought in Australia.

Flooding in San Diego, U.S.

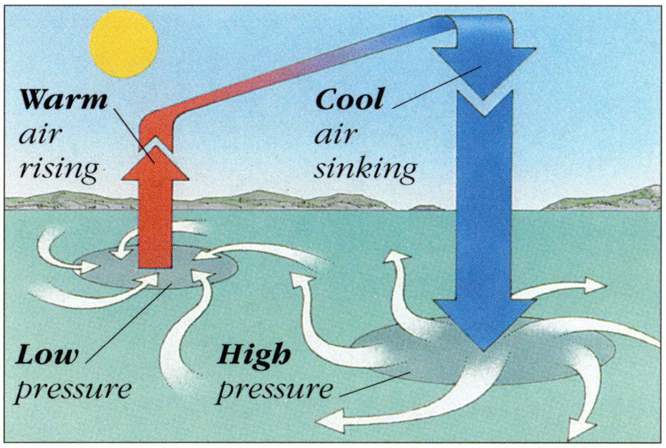

HIGH AND LOW PRESSURE

Warm air rises, making low air pressure areas. Cold air is heavier and sinks down, making high air pressure areas. Winds blow from high pressure to low pressure areas. Low pressure areas usually bring wet and windy weather, while high pressure areas bring dry and settled weather.

AMAZING FACTS

★ **The amount of water on the Earth stays the same and never changes. It just moves around and around through the water cycle.**

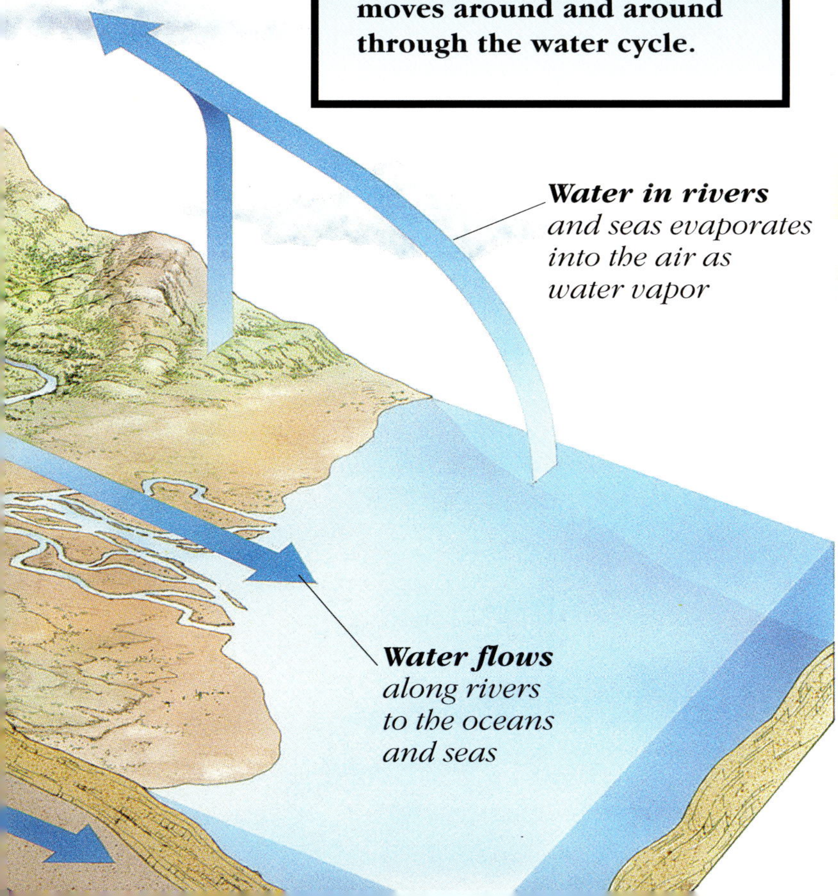

FIND OUT MORE
SCIENCE ALL AROUND US: Evaporation
TRANSPORT: Sailing ships

Forecasting the Weather

A weather forecast tells people what the weather is going to be like. Computers work out the forecast from measurements of rain, wind, air pressure, and temperature, and from information about what the weather has been like in the past.

The Earth's weather is watched from land, sea, and sky, as well as by satellites in Space. Weather satellites send pictures of cloud patterns and can measure temperatures all over the Earth. Large weather features, such as hurricanes, show up clearly on satellite pictures.

Photograph of Hurricane Gladys taken from a satellite in Space

COLLECTING INFORMATION

Information about the changing weather comes from thousands of weather stations on land, ships and weather buoys at sea, and aircraft and weather balloons in the sky. Computers combine this with information from weather satellites in Space to make weather forecasts.

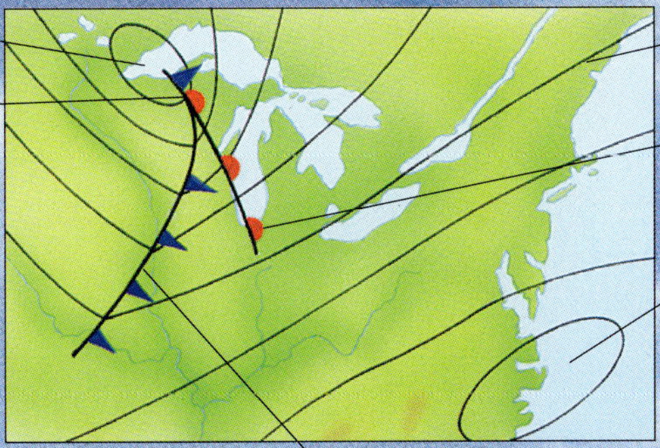

Weather map

Special planes *and ordinary aircraft record the weather high in the Earth's troposphere*

PICTURING THE WEATHER

The details of a weather forecast can be drawn as a special map. Lines on the map, called isobars, join up places with the same air pressure. Other symbols show features such as weather fronts, rain, sunshine, and clouds.

Weather balloons carry *instruments to measure temperature, wind speed, and air pressure in the troposphere and stratosphere*

Weather buoys collect *information from the surface of the ocean*

Gathering information about the weather

WEATHER MEASURING INSTRUMENTS

At a weather station, instruments measure the weather. A barograph measures air pressure, a thermometer measures temperature, a rain gauge measures rainfall, and an anemometer measures wind speed.

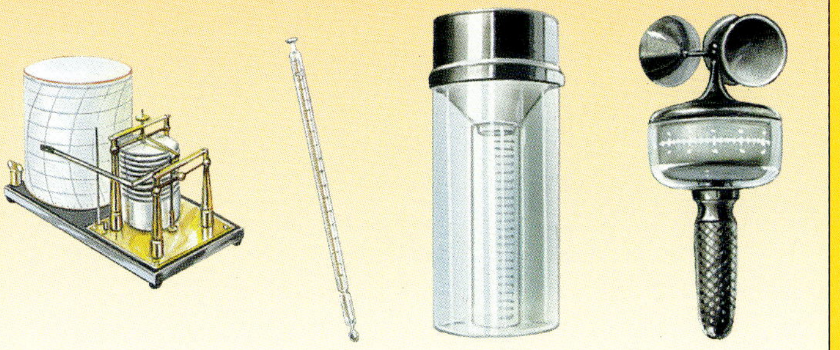

Barograph **Thermometer** **Rain gauge** **Anemometer**

FIND OUT MORE
HUMAN BODY: Thermometer
SPACE: Satellites

Clouds and Fronts

Clouds are made of tiny droplets of water or ice that float in the air. They form when warm air rises and cools down. This usually happens at a weather front, where a mass of warm air bumps into a mass of cold air. The warm air is lighter and always moves up above the heavy cold air. Along a weather front, there are often clouds, rain, and storms.

CLOUD TYPES

Clouds can be divided into 10 types, according to their height above ground level. The four main types are: cirrus clouds, which are thin and wispy; cumulonimbus clouds, which rise like towers; stratus clouds, which lie in layers; and cumulus clouds, which form in fluffy heaps with flat bottoms.

AMAZING FACTS

★ **About half the sky all over the world is covered with cloud at any one time.**

HOW A CLOUD FORMS OVER LAND

1 **The warmth of the** Earth heats the air directly above it, forming a bubble of warm air.

2 **The bubble of warm** air rises like a hot-air balloon, spreads out, and cools down.

3 **As the rising bubble of** air cools down, moisture in the air turns into drops of water, which make a cloud.

COLD AND WARM FRONTS

At a cold front, a cold air mass slides under a warm air mass. At a warm front, a warm air mass slides up over a cold air mass. At both kinds of front, warm air rises and cools, forming clouds and rain. A weather front can be hundreds of kilometers long.

A FOGGY DAY

Fog is cloud that forms near the surface of the Earth. It forms when a deep layer of air at ground level is cooled down by the land beneath it. Water vapor in the air condenses, or turns into droplets of water. These tiny droplets hang in the air as fog.

Truck driving through daytime fog

FIND OUT MORE
SCIENCE ALL AROUND US: Convection
TRANSPORT: Balloons

Rain and Snow

Rain and snow fall to the Earth from clouds. This happens when drops of water or tiny bits of ice in the clouds grow too big and heavy to hang in the air. The water or ice is pulled down to the ground by the Earth's gravity. Rain falls if the air is warm, and snow or hail fall if the air is very cold. Sometimes snow falls from clouds, then warms up and melts into rain on its way down.

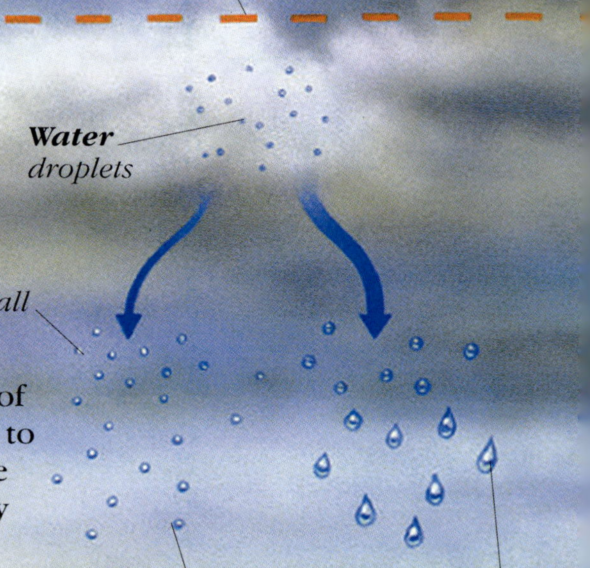

How rain and snow form

WATER OR ICE?

In warm areas, lots of drops of water in a cloud may join up to make bigger drops. When the drops are heavy enough, they fall to the Earth as rain. In cool areas, drops of water may freeze into ice crystals that stick together to make snowflakes, or collect more water and freeze again to make hail.

Measuring rainfall by collecting it in a jar

AMAZING FACTS

★ **The biggest hailstone recorded so far weighed 1 kg (2¼ lb). It fell in Bangladesh.**

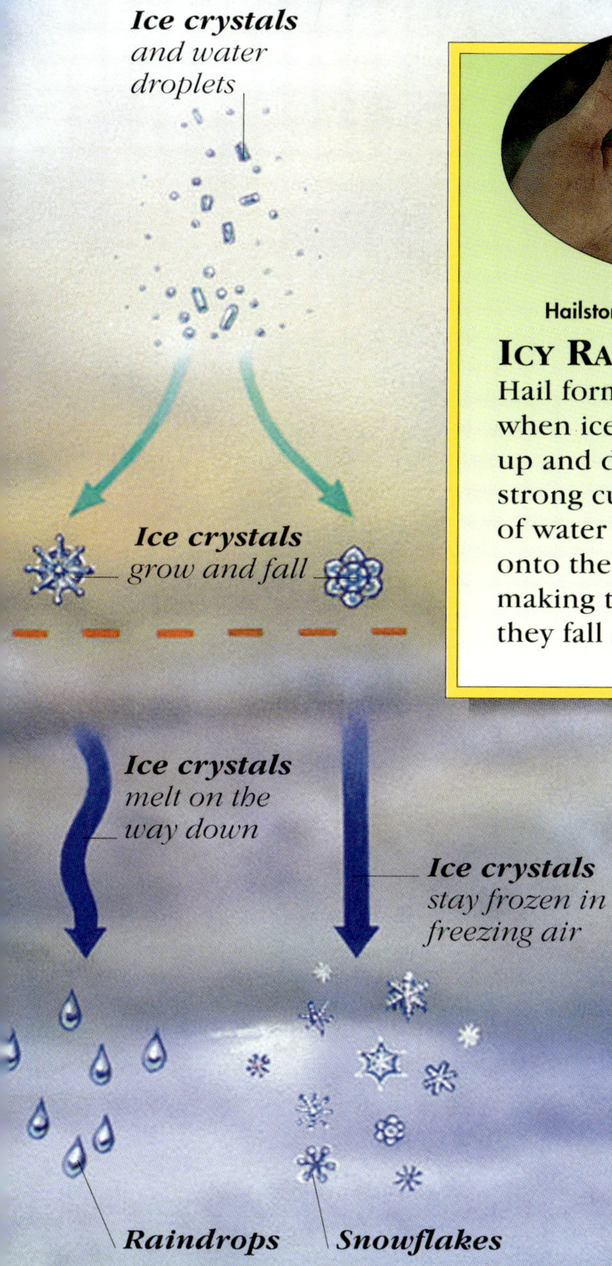

Hailstones in the palm of a hand

Icy Rain

Hail forms in a thundercloud when ice crystals are blown up and down inside it by strong currents of air. Drops of water in the cloud freeze onto the ice crystals in layers, making them bigger until they fall as hailstones.

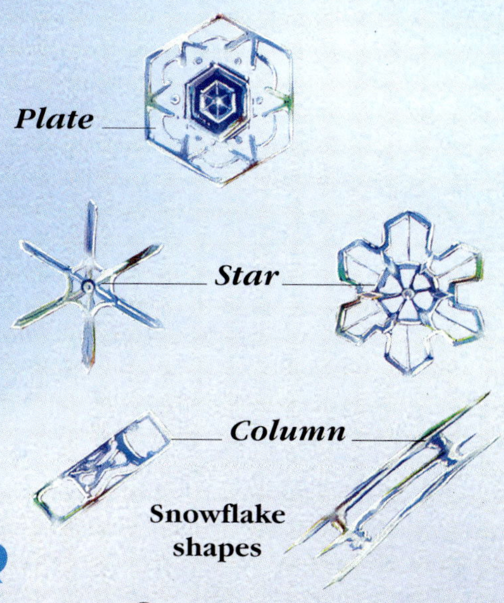

Crystal Shapes

Snowflakes can be grouped into plate, star, or column shapes. Their shape and size depends on the temperature, the amount of water in the cloud, and the height at which they form. No two snowflakes are ever exactly the same shape.

Flooded street after monsoon rain, Nepal

After a Monsoon

The heavy rain that happens in some tropical countries, such as India and Nepal, is caused by monsoon winds. These winds blow from opposite directions at different times of year. When monsoon winds blow off the sea and are full of water, they bring rain and floods.

FIND OUT MORE
CHILDREN OF THE WORLD: Monsoons
SPORT: Skiing

Stormy Skies

A storm is a time of bad weather, often with black clouds, heavy rain or hail, thunder and lightning, and a very strong wind. In cooler areas of the world, storms usually happen along weather fronts. In hotter areas, storms develop because hot air rises and cools to make clouds and rain. The worst storms are hurricanes, which form over the sea, and tornadoes, which develop over land.

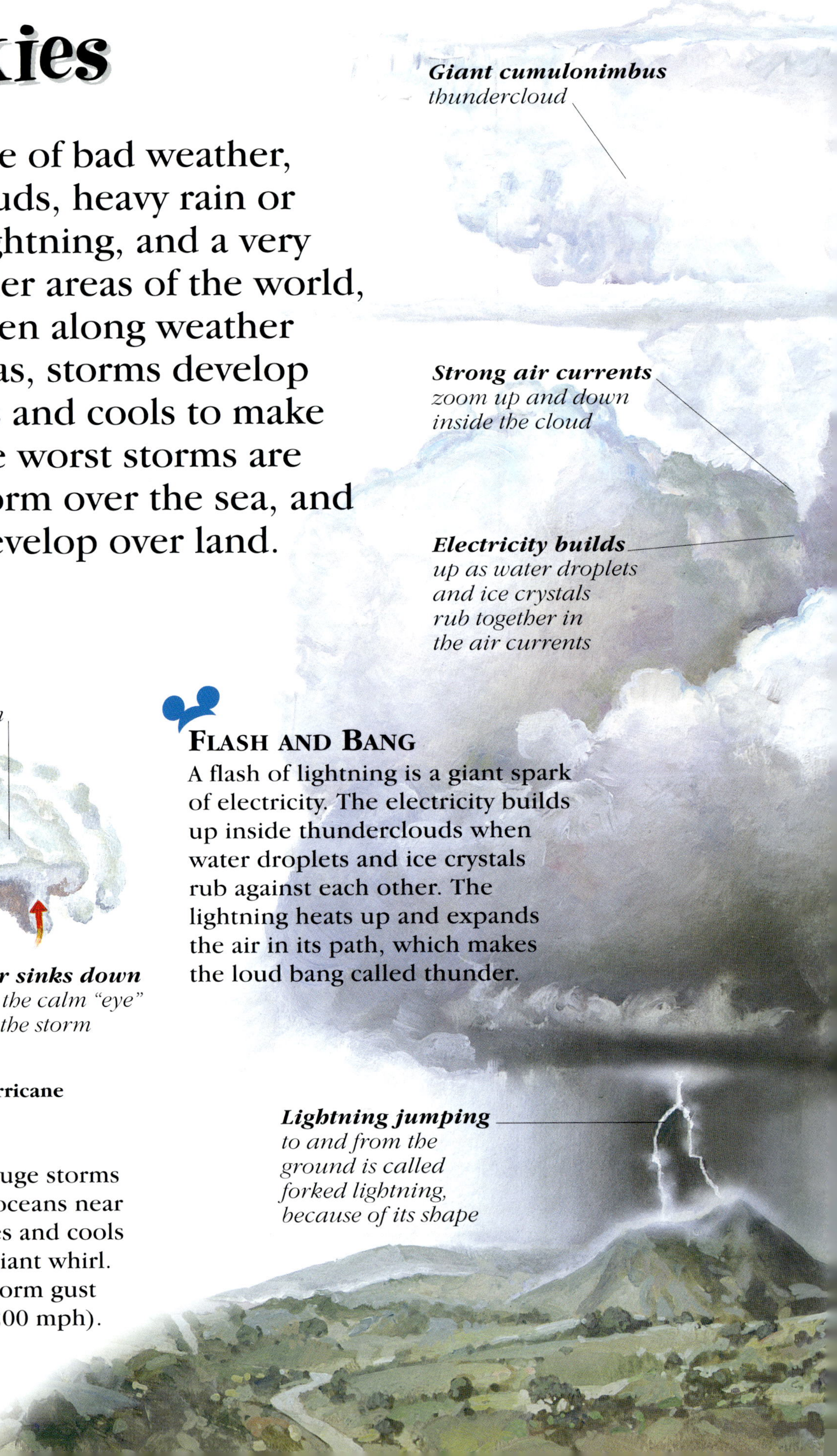

FLASH AND BANG

A flash of lightning is a giant spark of electricity. The electricity builds up inside thunderclouds when water droplets and ice crystals rub against each other. The lightning heats up and expands the air in its path, which makes the loud bang called thunder.

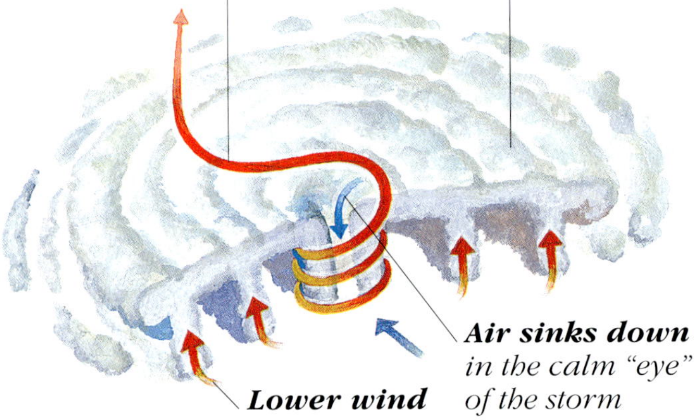

Cross-section through a hurricane

GIANT WHIRL

Hurricanes, or typhoons, are huge storms that form over warm, tropical oceans near the Equator. Hot, moist air rises and cools to make clouds that spin in a giant whirl. Winds around the eye of the storm gust to speeds of up to 320 km/h (200 mph).

Thunderstorm forming from a cumulonimbus cloud

Sheet lightning *occurs inside a cloud*

Lightning can strike *from cloud to cloud, as well as from cloud to air*

SPINNING COLUMN

A tornado is a spinning column of air reaching out from the base of a thundercloud. Air in a tornado is set spinning by strong winds in the top part of the cloud. The spinning speed of a tornado may reach up to 480 km/h (300 mph).

Tornado spinning over dry land

MEASURING THE WIND

The Beaufort scale describes the effect of wind speeds on people and surroundings. It is split into 13 levels called forces, from calm at force 0 to hurricane force at force 12.

Force 1 **Light air** **1–5 km/h (1–3 mph)**	Smoke drifts and leaves rustle	
Force 4 **Moderate breeze** **21–29 km/h (13–18 mph)**	Flags flap and dust blows about	
Force 8 **Gale** **62–74 km/h (39–46 mph)**	Twigs snap off trees	
Force 12 **Hurricane force** **118 km/h (73 mph) or more**	Widespread damage to buildings	

AMAZING FACTS

★ Every day, about 50,000 thunderstorms happen around the world.

★ Thunder occurs on about 220 days of the year in Java, Indonesia.

FIND OUT MORE
GREAT LIVES: Benjamin Franklin
SCIENCE ALL AROUND US: Electricity

Climates of the World

A place's climate is its pattern of weather from year to year. Climates are usually hot near the Equator, cold near the poles, and warm in between. Climates are affected by other things, too, such as mountains, winds, and ocean currents.

The Earth's climate has been changing for millions of years and is still changing today. This happens because of things such as changing amounts of heat coming from the Sun, huge volcanic eruptions, or pollution caused by people. In the past, there were periods of time called ice ages, when the climate cooled down.

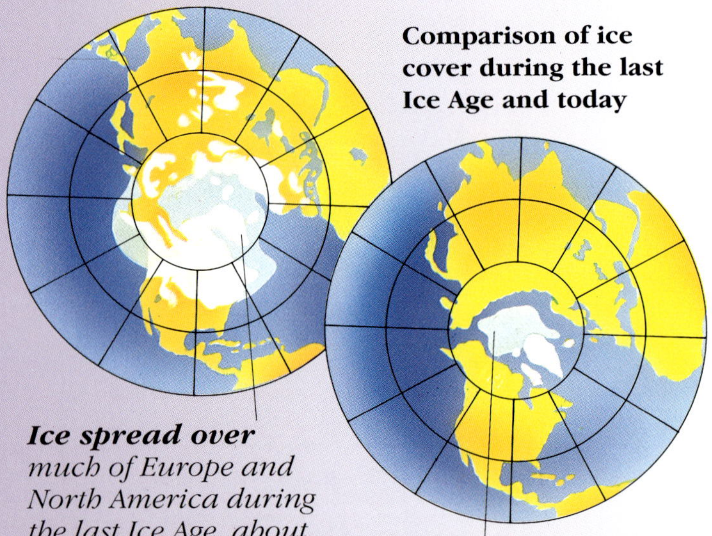

Comparison of ice cover during the last Ice Age and today

Ice spread over *much of Europe and North America during the last Ice Age, about 20,000 years ago*

Today, the climate *is warmer and there is less ice*

World Climate Zones

The world's climate can be divided into six main types: tropical, subtropical, warm temperate, cool temperate, desert, and mountain and polar. These can be separated into zones, or large bands, on the world map.

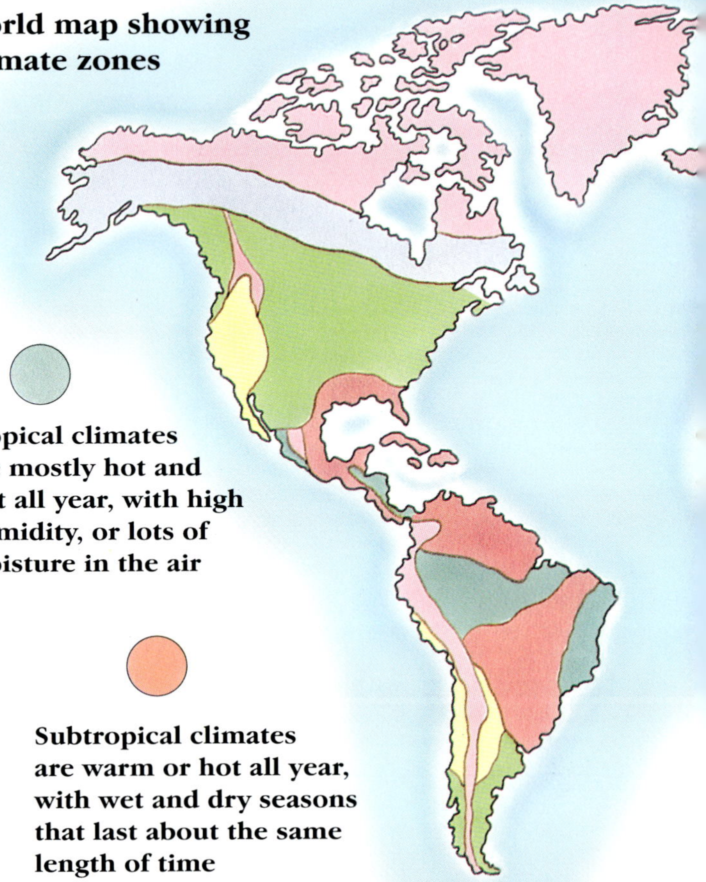

World map showing climate zones

Tropical climates are mostly hot and wet all year, with high humidity, or lots of moisture in the air

Subtropical climates are warm or hot all year, with wet and dry seasons that last about the same length of time

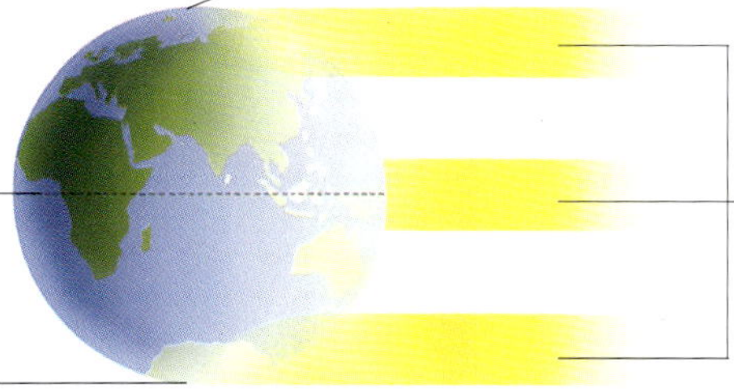

How the Sun heats the Earth

Warm temperate climates have four seasons, with hot, dry summers, cool, wet winters, a fall, and a spring

Mountain and polar climates are usually dry, with long, cold winters and short, warm summers

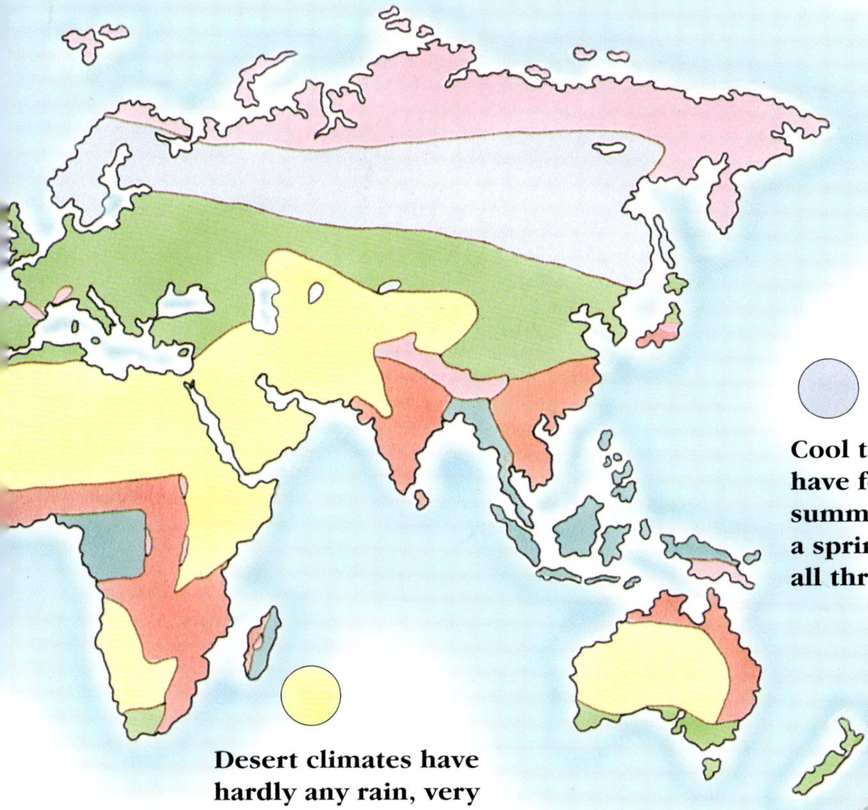

Cool temperate climates have four seasons, with warm summers, long, cold winters, a spring, and a fall. Rain falls all through the year

Desert climates have hardly any rain, very hot days, and very cold nights

CURVED EARTH

Different parts of the Earth receive different amounts of heat from the Sun. At the Equator, the Sun's rays heat a small area, making regions there very hot. At the poles, the Sun's rays spread out over a bigger area, making these regions cold. Areas in between have less extreme climates.

MOUNTAIN CLIMATES

The weather on a mountain is often wet on one side and dry on the other. As air rises over a mountain, it cools down to form clouds from which rain or snow fall. The air that flows down the other side of the mountain is dry and warms up as it goes down.

Climate of a mountain

FIND OUT MORE
ATLAS OF THE WORLD: World map
SPACE: Earth

Poles and Tundra

The North and South Poles are frozen, icy places at the top and bottom of the Earth. The area around the North Pole is called the Arctic. It is a frozen ocean with cold, flat land called tundra around the edge. Around the South Pole is a frozen continent called Antarctica. Both the Arctic and Antarctica have short summers and long, dark winters.

Wildlife of the polar Arctic and tundra regions

NORTHERN ICE

The polar Arctic scenery is always changing because sea water moves under the ice, making it crack and split. The ice is only about 4 m (13 ft) thick. Polar bears are the only land animals that live on the ice, but there are seals, whales, and smaller creatures in the sea.

ANTARCTICA

The continent of Antarctica is a frozen desert. It has mountains, volcanoes, and glaciers. Only animals that can survive the cold climate live or visit here.

Feathers and fat keep penguins warm in freezing Antarctica

AMAZING FACTS

★ Antarctica is nearly twice the size of Australia. It is covered with a layer of ice up to 4 km ($2\frac{1}{2}$ miles) thick.

Cross-section of tundra ground in summer

FROZEN EARTH

A little way below its surface, tundra ground stays frozen all year round. This frozen layer is called permafrost. In some places, the thin layer of soil at the surface thaws in summer. The ground becomes muddy and water collects in dips.

LIVING IN THE NORTH

The tundra lands of the Arctic are the most northern parts of Asia, Europe, and North America. They are so cold and windy that only a few plants, and animals such as musk oxen, can survive there.

FIND OUT MORE
ATLAS OF THE WORLD: Polar regions
MAMMALS: Arctic

Forests

Forests are areas where lots of trees grow close together. There are three main kinds of forests: tropical rain forest, coniferous forest, and deciduous forest. Forests give food and shelter to plants and animals. The trees' roots hold the soil together, and their leaves give off oxygen, which all living things need to survive. People cut down trees for their wood and clear the land for farming and building.

MADE FOR THE COLD

Coniferous forests grow in cold parts of the world, such as northern parts of Asia, Europe, and North America. Their trees have seed-bearing cones and leaves like needles that stay all year round. These evergreen trees are good at surviving the bitter cold of long, snowy winters.

Life in a coniferous forest

WHERE FORESTS GROW

Forest types vary with local climate. Tropical rain forests prefer hot regions and may grow on lowlands or at the foot of mountains near the Equator. Deciduous trees prefer cooler regions, sometimes on the slopes of mountains. Conifers grow well in cold places nearer the poles or high up mountains. Some places are too cold for any trees to grow.

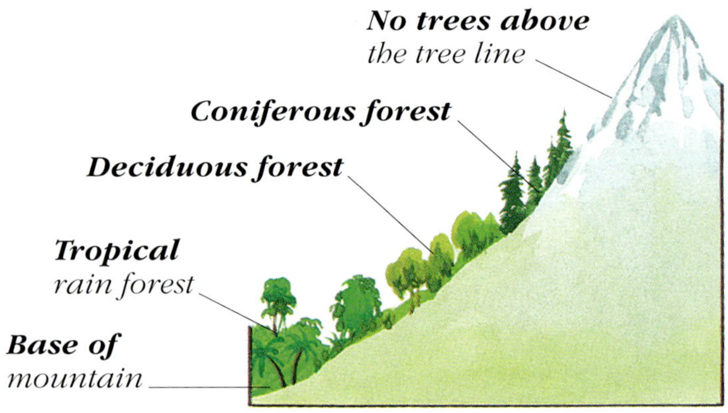

Forests and climate

TREES FOR ALL SEASONS

Deciduous forests grow in areas of the world that have seasons, such as parts of China, Japan, Europe, and North America. Deciduous trees shed, or lose, their leaves just before winter or a dry season. By shedding their leaves the trees can survive cold and drought.

Life in a deciduous forest

TROPICAL TREES

More types of plants and animals live in a rain forest than in any other kind of habitat in the world. It is hot and wet all year round, so the trees grow steadily. Very little light gets through to the forest floor. Most animals and plants live high in the treetops. Soil is easily washed away if the trees are cut down.

Life in a rain forest

AMAZING FACTS

★ There may be as many as 950 different kinds of beetles living on just one rain forest tree.

FIND OUT MORE

MAMMALS: Rain forests
PLANT LIFE: Trees

Grasslands

Grasslands are large areas of flat or gently rolling land covered by grasses. There are few trees. Grasses grow well in these open lands, even when damaged by winds, droughts, fires, or grazing animals. Many types of grasses grow all over the world, and grasslands have different names, such as savanna, steppe, pampas, or prairie, depending on where they are. They are rich in wildlife.

Grazing animals on the Asian steppe

STEPPE

The temperate, dry grasslands of central Asia and Eastern Europe are called steppe. In summer the temperature can reach 30°C (86°F), and the winter can be as cold as -30°C (-22°F). Decades ago, herds of wild horses used to eat the short grasses. Today, farmers grow wheat and graze sheep on the steppe.

THE CYCLE OF LIFE

On grasslands, no food is wasted. For example, on the African savanna, herbivores eat the grass. They are hunted by carnivores or scavenged by other meat-eaters. Any remains of dead animals are broken down by insects and bacteria, and nutrients are returned to the soil.

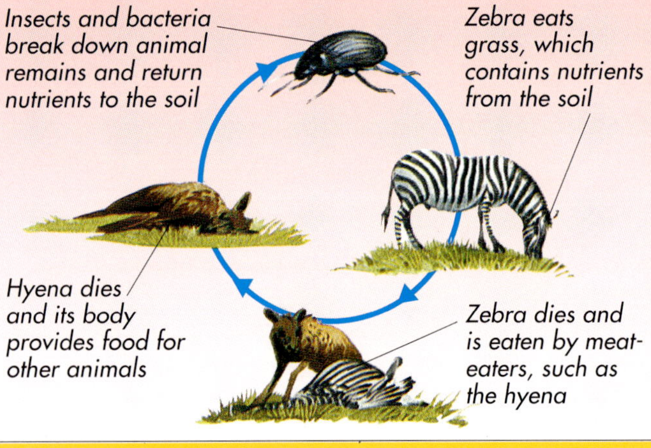

Ground level on the African savanna

SAVANNA

The grasslands of tropical Africa are called savanna. The weather is hot all year round, but there are wet and dry seasons. The grasses in the savanna are either high grasses, up to 4.5 m (15 ft) tall, or short grasses, about 30 cm (12 in) tall. Animals such as wildebeest and giraffe move across the grasslands following the rains.

Many animals live on the pampas of South America

Amazing Facts

★ The world has about 10,000 different kinds of grasses.

★ Thorny bamboos in India grow to 37 m (120 ft) and are the tallest grasses in the world.

Pampas and Prairie

On the wetter, milder grasslands of the pampas in South America and the prairies in North America, the grass can grow tall and green. The pampas is home to many kinds of burrowing animals, such as cavies, seed-eating birds, such as rheas, and termites that build and live in tall mounds of soil.

Small grassland *animals, such as the ground squirrel, shelter in burrows*

Larger animals, such as *giraffes, graze on the leaves of the few trees and bushes, and drink at water holes*

Prairie fire, North Dakota, U.S.

Burning Grasslands

During dry seasons, fires are often started in grasslands by lightning or people. Fire destroys young trees and bushes, which prevents the growth of woodland and forest. After a fire, however, the grass grows back fresh and strong.

FIND OUT MORE
CHILDREN OF THE WORLD: Gauchos
MAMMALS: Savanna

Deserts

Deserts are dry places where there is little or no rainfall. Most deserts are very hot during the day and very cold during the night. Some deserts are cold all the time. Deserts sometimes have sandy soil, but most deserts are covered in stones, gravel, and bare rock. Few plants survive in the dry desert soil. Rocks are exposed to wind and rain and are worn into strange shapes.

AMAZING FACTS

★ Deserts cover about one-third of the Earth's land surface.

★ The world's highest sand dunes, in the Sahara Desert, are 465 m (1,525 ft) high.

★ The Atacama Desert in South America had no rain for 400 years.

Family in the Gobi Desert, Asia

Desert landscapes

DESERT SURVIVAL

The Gobi Desert in central Asia has freezing cold winters, when temperatures fall to -40°C (-40°F). Even in summer, cold nights follow the hot days. Rainfall in deserts is so rare that plants and animals there have adapted to survive with little water. Most of the Gobi Desert receives less than 25 cm (10 in) of rain a year.

DESERT LANDSCAPES

Deserts are mostly rocky places, with strange shapes carved out of rocks by wind-blown sand. Desert winds also blow the sand into heaps called dunes. Sometimes storms happen, and the heavy rain quickly carves out valleys and wears away softer rock.

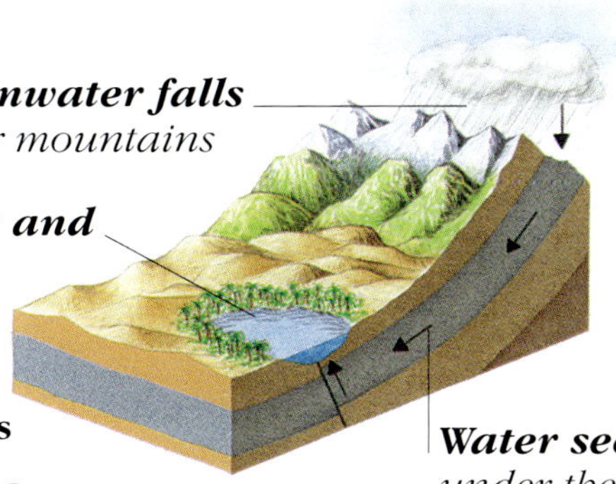

WATER IN THE DESERT

An oasis is a place where water is found in a desert. Water collects in hollows worn out by the wind. It may come from desert storms or from rain that falls on mountains on the edge of the desert. Rainwater drains through rocks deep under the desert and collects at an oasis.

FIND OUT MORE
MAMMALS: Gobi Desert
PLANT LIFE: Desert plants

Helping the Earth

People's activities are changing the natural balance of the Earth. They are changing the weather, using up the Earth's resources, and destroying wildlife habitats. To help the Earth and its living creatures survive, people must produce less pollution, use less energy and resources, and recycle more of the things they make.

Ozone Holes

A layer of ozone gas high up in the Earth's stratosphere keeps out many of the Sun's rays. Some of these rays can do harm if they reach the Earth. They can cause people to develop skin complaints. Certain chemicals released from aerosols and refrigerators are making holes in the ozone layer, especially over the North and South Poles.

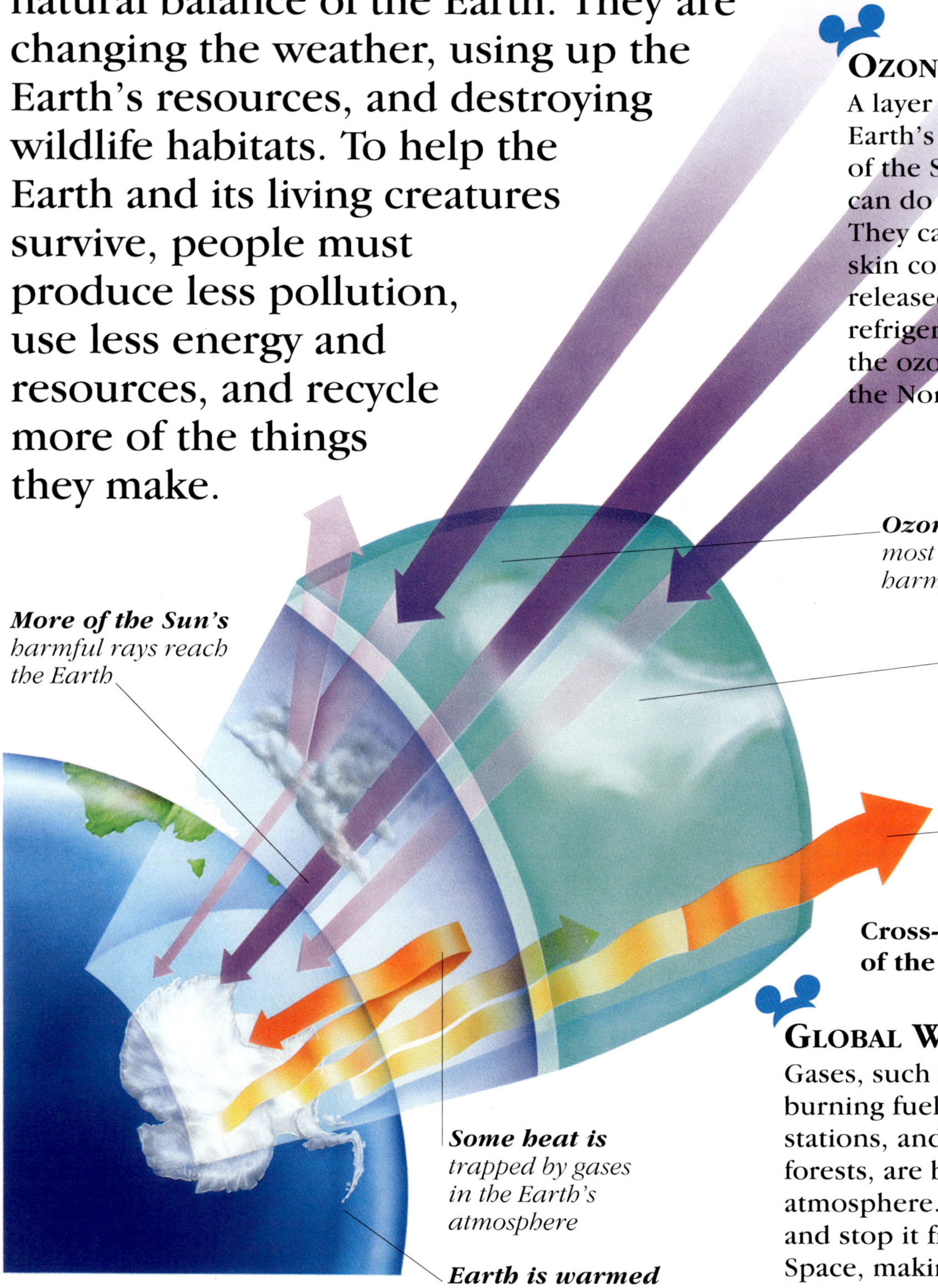

Cross-section of part of the Earth's atmosphere

Global Warming

Gases, such as carbon dioxide from burning fuels in factories, power stations, and vehicles, and burning forests, are building up in the Earth's atmosphere. They trap the Sun's heat and stop it from escaping back into Space, making the Earth warm up a little. This is called global warming.

POISONOUS RAIN

Harmful gases, such as sulfur dioxide and nitrogen oxides, from vehicles and factories mix with clouds and make rain more acidic than normal. Acid rain harms wildlife and trees, and eats away at stone buildings. Reducing pollution from cars and factories helps prevent acid rain.

Acid rain pollutes lakes and rivers

Replanting a forest in South America

NEW TREES FOR OLD

People are cutting down too many forests for wood or to make way for farms and cities. In some forests, new trees are planted to replace those cut down, making sure there will still be forests on Earth in the future. Forests are important because they affect the weather, protect the soil, and give plants and animals food, shelter, and oxygen.

RECYCLING

Paper, glass, metal, fabric, and plastic can all be recycled or used again. This helps use fewer new materials to make things. It also means there is less waste to fill up garbage dumps, which spoil the countryside and pollute the air, water, and soil.

Trash can be sorted into types of materials for recycling

AMAZING FACTS

★ The sea level is expected to rise by 15–46 cm (6–18 in) in the next 50 years because of global warming.

FIND OUT MORE
GREAT INVENTIONS: Recycling
PLANT LIFE: Managing plants

Glossary of Key Words

Acid rain: Rain mixed with gases from the air, which make it harmful to wildlife and their surroundings.

Atmosphere: The gas that surrounds a star, planet, or moon.

Caldera: The round, shallow basin formed when the top of a volcano blows off in an eruption.

Cirrus cloud: Thin, wispy clouds made of ice crystals.

Climate: The usual pattern of weather in an area over a long time.

Continent: One of the seven large landmasses into which the Earth's land is divided. The seven continents are Asia, Australia, North America, South America, Europe, Africa, and Antarctica.

Core: The central part of the Earth. It has two layers: the inner core made up of solid metals, and the outer core made up of liquid metals.

Crust: The thin "skin" of rock that covers the Earth's surface.

Cumulonimbus cloud: A huge storm cloud that brings heavy rain, thunder, and lightning.

Cumulus cloud: A fluffy, white cloud that usually means good weather.

Current: A flow of air or water moving in a certain direction.

Delta: A flat area of sediment dropped by a river when it flows into a lake or sea.

Earthquake: A shaking of the ground caused by a sudden release of energy stored in the rocks under the Earth's surface.

Equator: The imaginary line that runs around the middle of the Earth, dividing it into two.

Erosion: The gradual wearing away and removal of rock and soil by water, wind, or ice.

Fault: A crack or line of weakness in the rocks of the Earth's crust.

Fossil: The remains of a plant or animal preserved as rock. Imprints in rock, such as footprints or skin, can also become fossilized.

Front: Weather changes that occur when a warm air mass and a cold air mass come together.

Geyser: Jets of hot water or steam that shoot up from hot springs beneath the ground.

Glacier: A river of ice that moves slowly down a valley.

Global warming: The gradual warming of Earth's atmosphere due to the build up of certain gases such as carbon dioxide.

Gravity: The force that pulls everything toward the center of a star, planet, or moon.

Hemisphere: One half of a globe or sphere. The Earth is divided by the Equator into the northern and the southern hemispheres.

Hurricane: A violent tropical storm that forms over warm seas near, but not at, the Equator.

Igneous rock: A rock formed when magma or lava cool down and harden.

Lava: The very hot, liquid rock that erupts from a volcano, and then cools and hardens on the surface of the Earth.

Magma: Hot, melted rock in the Earth's mantle that sometimes escapes through volcanoes.

Mantle: The deep, heavy rock inside a planet, such as Earth, on which the thin surface crust floats.

Metamorphic rock: A rock formed when existing rocks are changed by heat or pressure.

Mineral: One of the non-living substances found in the Earth's crust. Some minerals are mined and used to make other materials.

Monsoon: A wind that blows from different directions at different times of the year, causing wet and dry seasons.

Moraine: Collections of rock pieces eroded by a glacier that form a ridge on the glacier, or on the ground after it has melted.

Oasis: An area in a desert around a well or spring. An oasis often has enough water to support plant life.

Ozone layer: A layer of ozone gas in the Earth's atmosphere that absorbs harmful ultraviolet rays from the Sun.

Permafrost: In polar regions, a layer of soil below the surface that stays frozen all year round.

Sediment: Tiny pieces of rock or the remains of living things that settle in water.

Sedimentary rock: A rock formed when layers of sediment are pressed together.

Stratus cloud: A grey, shapeless layer of cloud that often gives long periods of drizzle.

Trade winds: The name given to the winds that blow steadily toward the Equator.

Tsunami: A giant ocean wave triggered by an undersea earthquake.

Tundra: A vast, flat, treeless plain in Arctic regions. The ground is often frozen.

Index

(*see* **Famous Places** for a full index to your complete set of books)